I0834640

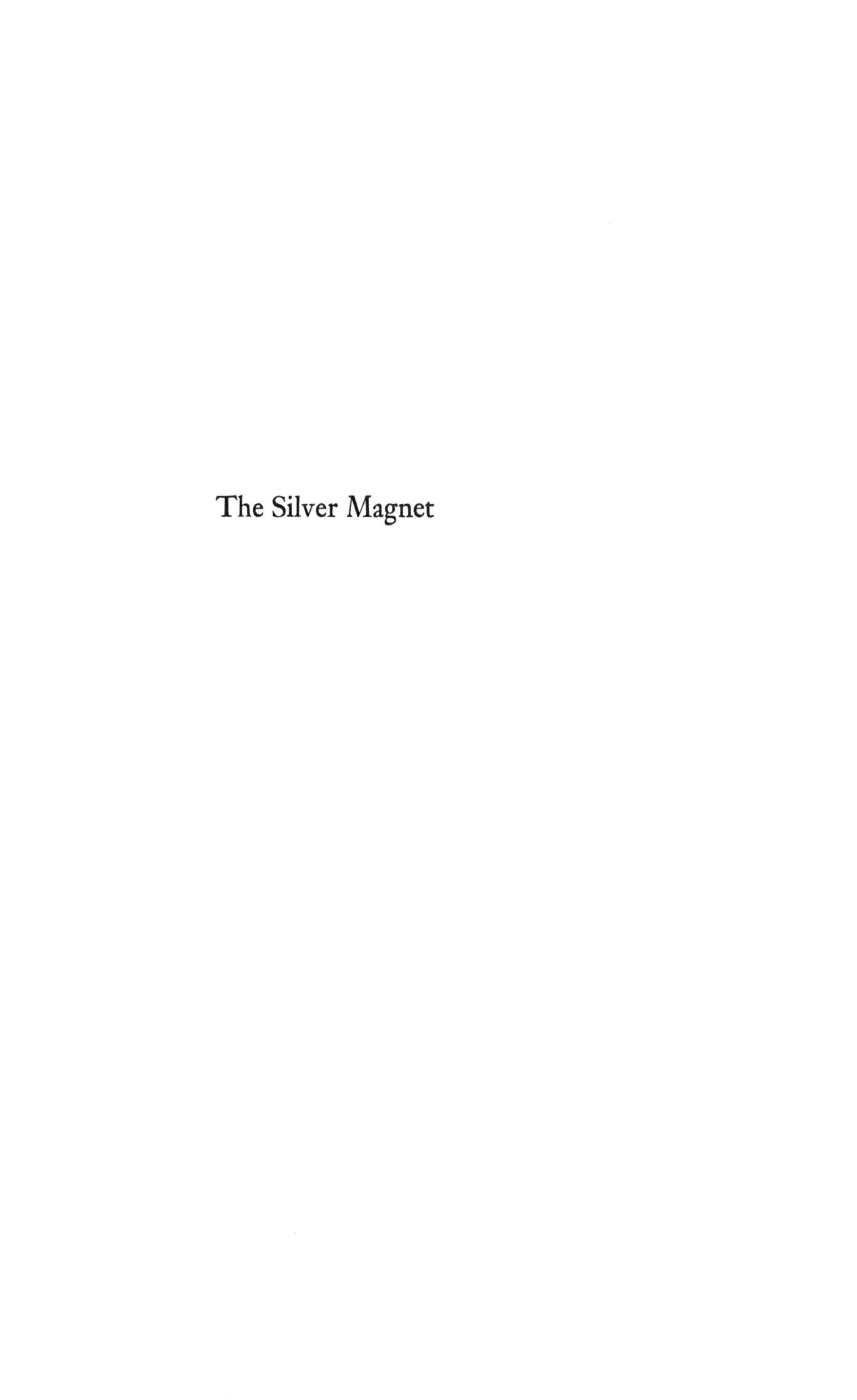

The Silver Magnet

The SILVER MAGNET

FIFTY YEARS IN A MEXICAN SILVER MINE

GRANT SHEPHERD

Illustrated

New York E. P. DUTTON & CO., INC. *Publishers*

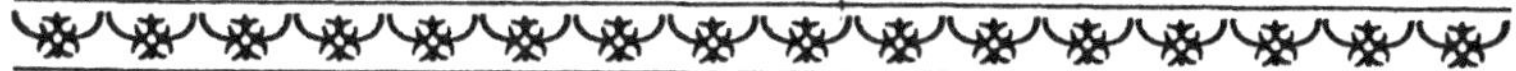

FIRST EDITION

PRINTED IN THE UNITED STATES OF AMERICA
BY THE WILLIAM BYRD PRESS, INC.
RICHMOND, VIRGINIA

*To my beloved and long suffering wife,
whose knowledge of punctuation and spelling has
made this book possible*

Contents

Illustrations

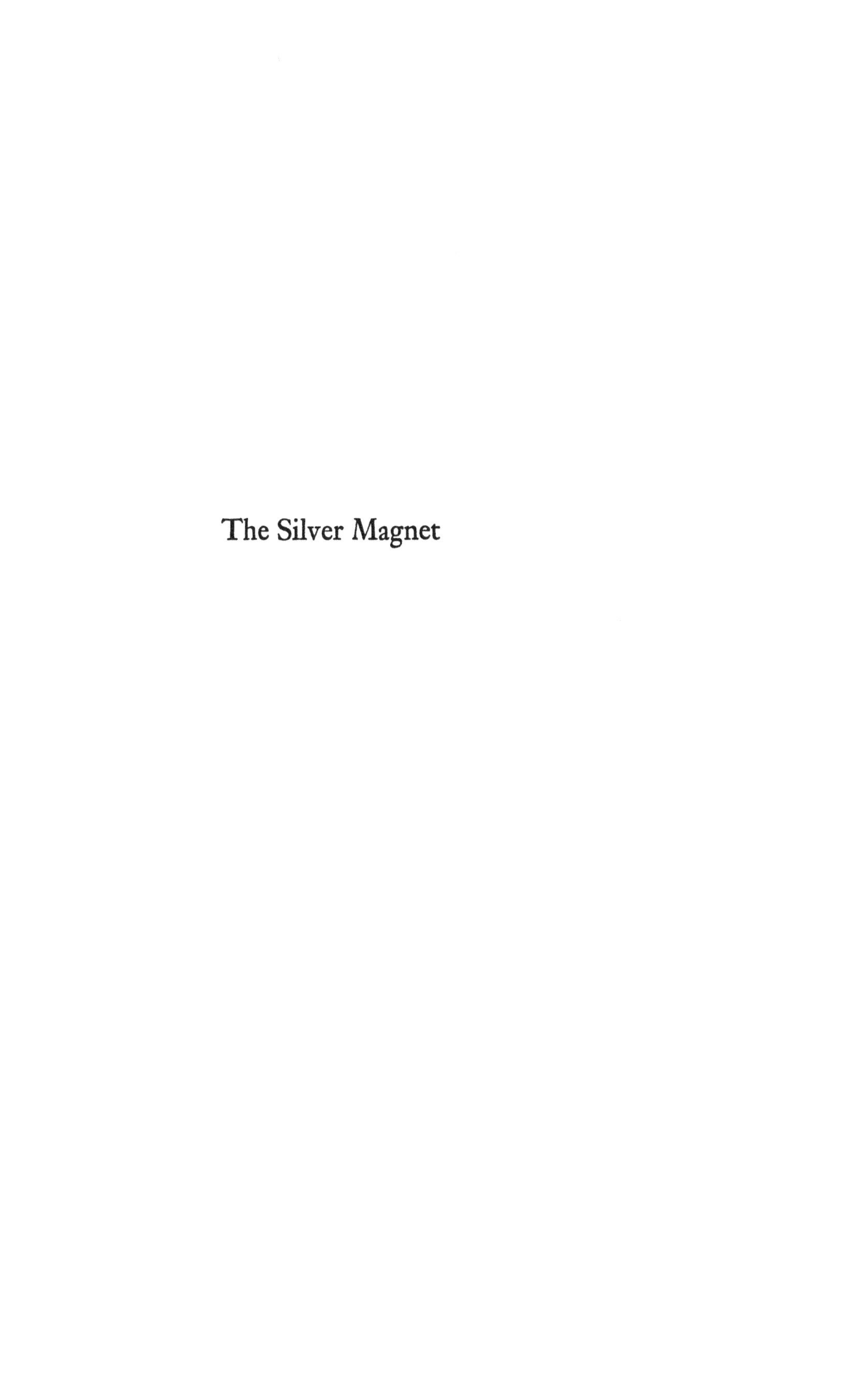

The Silver Magnet

I

Departure

In the year of our Lord 1880, my father, Alexander Robey Shepherd, erstwhile Governor of the District of Columbia, and his lovely, brave and devoted wife, with their children, servants, dogs, employees and friends, packed up bag and baggage in Washington, D. C., and set out for the ancient village of Batopilas in the State of Chihuahua in the country of Mexico.

To this modern generation such a journey presents no features of especial discomfort. But fifty-seven years ago it took forty days . . . this trip into that little known and as little traveled country. It was made feasible by a letter to the Commanders of the Army Posts throughout the State of Texas from General Sherman, who was my father's very good friend; the Baltimore & Ohio Railroad and others; ambulances, wagons, horses, mules—hûman and otherwise.

The first Army Post in Texas where we rested for a few days during this adventurous period of travel was Fort Concho. There I experienced the first great fright of my life. I was perfectly terrified despite the fact that I fought, like all small boys do, with my young brother, and only within the last half hour had been pulled from off him by my long-suffering nurse. But it seemed to me at the age of five years that he was in imminent danger of being trampled to death beneath the feet of marching troops, and it threw a scare into me that makes this particular incident of those long-past days vividly clear. It is really much clearer than many other and far more important occurrences of that memorable trip.

Of course, he was not trampled to death. Those whom I had regarded as ferocious soldiers opened their ranks very slightly, and the Kid passed through them unharmed, marching in time to the drum with his little stick on his little shoulder entirely oblivious of the fact that he had caused considerable consternation to his family who were admiring the parading troops on the parade ground. Oblivious, yes. Until his nurse, a red-cheeked, buxom Irish girl landed on him and ignominiously swung this venturesome three-year-old into her arms with voluble and heavy-brogue remonstrances.

Mary Jones she was, and far from the shores of the Emerald Isle. There were many, many new things she had to learn, along with her employers, in this recent service, an early and very painful one being that the prickly pear *is not*, as she had fondly believed "sure the gr-r-andest, foinest Shamrock I ever-r-r saw!" Poor Mary Jones! And poor mistress who later had to pick out the thorns from the too eagerly grasping fingers.

There must have been considerable rivalry between the gallant Infantry at the various Army Posts through which we passed and the long-moustachioed teamsters over Mary's charms. But stops at Army Posts were brief at best. Mary as history records arrived unscathed with all the rest of us at her final destination, a then small ancient silver-mining camp in the Sierra Madre Mountains of Mexico. For many long years, with only an occasional ruction with my own much beloved German nurse, Dora, she helped to get that big family shaken down into the routine of a novel and entirely extraordinary environment.

Dora was a dear, faithful creature, devoted and loyal to the heads of the family, as well as to the seven less important members, during that long term of years. She saw them grow up, marry, bring forth other generations in their own good time. These others like their parents spoke Spanish, played and fought with the little Mexican children; rode burros when they were quite small and were promoted first to ponies and

General View of Hacienda San Miguel.

1. Refectory
2. Office Buildings
3. Corral and Stables
4. General Manager's Residence
5. Machine Shop
6. Foundry (Iron)
7. 25 Stamp Mill
8. Amalgamation Sheds
9. Refinery
10. Cyanide Works
11. Store
12. Assay Office
13. Lixiviation Works

Governor Alexander Robey Shepherd on "Sutton" in front of main office. Main gate in background.

Some of the "Gang." Con and Grant Shepherd, Louis Mohun, Frank English, Tom Shepherd, Gilbert Geiselhart, Ed Stevens, "Dick."

then to true saddle animals. Dora forgave me much and forgave it often for I was, in truth, a vexatious small boy.

But I have traveled a thousand miles or more in six or seven paragraphs.

Let us attempt to visualize the Baltimore & Ohio Railroad Station in Washington on the evening of May 1, 1880, at the hour of ten. It will be difficult to do this, for while the Baltimore & Ohio Station was located more or less on part of the same ground now occupied by the Union Station, there is a vast difference now between those two places.

The family consisted of seven children . . . May, about nineteen; Isabella who was eleven, with Susan and Grace at appropriate ages between those two; there were three young male animals, Alexander II aged eight; Grant, five; and John Conness aged three; their parents, their four dogs, their two nurses; and Uncle Jack. Uncle Jack was beloved by all of us; in his early twenties, fine-looking, and a good horseman. These made up the members of the family, a total of twelve. Aside from this family group the following gentlemen were in the party, who were going down to Batopilas as the American nucleus for the organization at the mines: Dr. Ross, surgeon; Colonel Morgan; L. M. Hoffman; Lyman Larned; and Mr. Lindsay, engineer. We were seventeen souls and it is readily understood that a special Pullman was needed and used for the railway part of the journey.

The general outline for this route was from Washington, D. C., to St. Louis, Missouri; and from that place to San Antonio, Texas, by rail. From San Antonio by wheeled vehicles and "hay burners" to Forts Concho, Stockton, Leon Holes, Davis; on to the last place in the United States of America, which was Presidio del Norte, then across the Rio Grande to the city of Chihuahua in Mexico. From this last-named city we were finally to set forth for our objective, Batopilas.

At the hour and date mentioned, many friends and relatives were assembled at the railroad station in Washington. There began the stowing away of innumerable bags, baskets, boxes of fruit, lunches and candy; four dogs . . . of which two were in crates. The other two dogs were too large for any crate, they were magnificent mastiffs. The crated canine sufferers were a Chesapeake Bay pup, a gift to my father from his friend General Babcock; and a setter pup presented by Tom Cropley.

I have been told that the parting was a sorrowful one. It was a big jump into the unknown; not a person in the party except my father had made the trip previously. While the journey made by my father in the year 1879 presented no particular difficulties to a man of determination and physical vigor like himself, that same trip was an undertaking indeed when it was repeated with a household such as I have described.

The lighting in the railroad coaches was furnished by coal-oil lamps of assorted sizes and construction. Therefore, it was easier than it would be now to disguise tears and blushes, although both of those peculiarities were more freely indulged in then than now. Gradually the Pullman filled up and the tearful good-byes began. The confusion somewhat subsided as the conductor gave his "All aboard!" Too-full throats were cleared, tears were transferred from cheek to handkerchief, and the train started to rock its way toward distant St. Louis. You might have thought that then all would be serene. But other ordeals still had to be gone through with . . . the putting to bed of the children, the locating of different pieces of hand luggage in which were night-shirts, and so forth. The children were tired out and must have been performing as all obstreperous tired children will and do perform on a railroad journey.

I quote from my mother's diary: "There is no time for sad thoughts, for tired and excited babies must be bundled into berths which they occupy for the first time in their short lives; notwithstanding which fact, they lie down without remark upon the novelty, save perhaps a tired 'isn't this funny?' and are soon asleep."

Possibly the man who wrote "Rocked in the Cradle of the Deep" received his inspiration from a sea voyage; he could have got just as vivid a picture by taking any railroad journey in those days. The fact that we suffered from car-sickness was more than pardonable, almost justifiable, though nevertheless decidedly unpleasant. My poor mother first, and then everyone, is entitled to our sympathy; in charity let that first night remain with no further description.

As I said before, my father in the year 1879 made the trip into Mexico, lured by the silver magnet, to investigate the possibilities of the ancient property known as Batopilas, which had been first discovered in 1632. He sailed from New York City down the Atlantic Ocean to Panama, and crossed the Isthmus on the railroad. He went up the Pacific Ocean northward in a sailing vessel to the Mexican port of Ajiabampo. From that place the hot coastal plain was crossed on the hurricane deck of a mule, and so into the Sierra Madre Mountains to the silver-mining camp of Batopilas. On the return trip back into the United States, he came out of Mexico northeast across the Sierra Madres, still on the hurricane deck of a mule, to the City of Chihuahua and thence into the United States exactly the same route you are going to take with me, but in reverse order.

It took us three long, dusty days to reach St. Louis. We children were sick many times as a result of too much candy, the Baltimore & Ohio Railroad bed, and so on. We made even more than the usual nuisances of ourselves. At St. Louis we transferred to another railroad and to yet another special Pullman and had opportunity to observe many unfortunate emigrants with their unwashed faces, their dirty bundles, their strong smells, who were on their way to settle that great Western country of ours.

The major portion of our movements from day to day is extremely hazy to me. I do recall very vaguely that upon arrival in Houston, Texas, the train onto which our Pullman was to have attached as tail-end had already left. There was no small verbal warfare, for my father was a forceful man. A special Pullman with a dozen or more passengers was something which ought to have been waited for, unless the train which was pulling it was so very late that it could go as well on tomorrow's schedule as on today's. . . .

After Houston, what? The Hotel Menger in San Antonio is the answer. In 1880 that hostelry was an unanswerable argument.

We bade good-bye to the hospitable people in San Antonio with reluctance, but not so with the Hotel Menger. There we also said good-bye to the "iron horse" and outfitted for the over-land-and-sand journey to the last halt in the United States, Presidio del Norte on the Rio Grande. To be a true 1880 Texan, you pronounced that river Rio Gran, with the emphasis on the "I" as in "eye."

Outfitting, that was some job of work!

Having grown up in an atmosphere of miners, teamsters, arrieros, in later years I could begin to appreciate what a task it really was. What a sweet time the natives of San Antonio, experts in the ancient and honorable profession of horse and mule barter, must have enjoyed with my father and my uncle and the other male tenderfeet in the party. Both my father and my uncle knew a good deal about blooded horses, but about a quantity of wagons with their six- and eight-mule teams making up a wagon-train, they were so very ignorant that they flinched in both front and hind feet.

There were several ambulances in the cavalcade. The family, the accompanying physician, and the other gentlemen in the party rode in this luxurious mode of conveyance. An ambulance is a heavy Dayton wagon, only more so. You who drive your five hundred miles a day over roads as smooth as a billard table in an automobile can have no conception of how your back feels after, say, fifty miles "cross country" in one of those vehicles. I expect that is one of the many reasons which gave those oldsters a stiff backbone and a small diameter at the equator.

The wagons with the supplies were given a ten-hour start of the ambulances from the town of San Antonio. But we overtook them before Leon Springs was reached. The first night in camp was passed in grave anxiety at that place, and two or three days besides, while the doctor labored to save the life of my mother, who had succumbed to the effects of the food at the Hotel Menger.

I am not describing any "covered wagon" episodes. Our travels for that time were trekking de luxe. In fact, I clearly remember one morning my father was tremendously incensed and more than probably decidedly profane because some of the teamsters had stolen an entire case of his best Port wine, and were, therefore, in no condition to start out. His ire, as was the case in Houston when we missed connection there, was not on this occasion so much at the loss of time entailed in contending with heavy-lidded drivers—who were absolutely poisoned by good Port—but because he had progressed far enough out "at sea" to appreciate the fact that the purveyor of all good things to eat and drink was a mighty long way to the north of him.

It was on that same morning while I was riding proudly on the driver's seat beside "Dolf," with his long-horn moustache, that I had the thrill of seeing him offhandedly shoot a great rattlesnake which was coiled near the roadside. Those roads consisted of a very occasional wagon track in the sand and mesquite filled with rocks of assorted sizes. One day my father was almost disjointed after miles of such travel, and he promised the driver a bottle of beer for every rock he would miss. I fear that the debt was never paid in beer. Those old Texans liked their liquor with "the thorns still in it" and at the rate of ten drinks to the full quart. Maybe he found, in some of those little towns and Army Posts through which we passed, enough whiskey to pay the debt. Any driver would have to miss a few of those rocks no matter how carelessly he drove.

The boys who have longed to be street-car drivers, or motormen, racers of automobiles, fliers, and so forth can imagine the yearning to become a "Dolf" with a long yellow moustache shaped exactly like the horns of a water buffalo and nearly as long. A "Dolf" with a lean, sun-tanned face, sitting up there having a six-shooter strapped to you, your hand full of the six lines attached to those ornery, jackrabbit-eared mules, and a hickory-stock whip with which to pop a horsefly from a mule at any time in the most nonchalant and fetching manner.

It will always be a mystery to me how buxom Mary Jones ever resisted him or the many others, in fact. I'm sure had I been a Mary Jones there would have been a different tale to tell. After the passage of all these years, I fail to recall whom it was that Mary selected as a mate, but I do remember that it was a successful marriage, and that the couple lived not far from El Paso, Texas.

A child in later years may recollect certain incidents, a few at any rate, of such a journey. It must be remembered that the lucky child when weariness overtakes him goes to sleep in some grown-up's lap or other comfortable place, the making available of which has necessitated the still greater discomfort and crowding of the grown-ups. They must for hours endure true aches and pains sitting long in cramped positions, while the patient, weary mules drag the heavy vehicles over miles and miles of bad roads, across a hot, dry, dusty country to a temporary camp at night; or to an occasional by comparison gloriously commodious Army Post, where every kindness and consideration possible is shown to the almost exhausted travelers.

Those camps must have been a tedious ordeal to the grown

people, particularly as the colored man-cook, picked up in San Antonio, turned out to be an incompetent and useless fellow. It was here that Dora's devotion to my mother and her seven children demonstrated that she was a true and godly Christian. I have, since those days, met many men and women called godly, and some of them were; but it is my firm conviction than when dear old Dora arrived at the pearly gates there was no delay in the opening of them, nor in her being escorted to her well-earned place of the greatly deserving.

Of the kindness and affectionate welcome accorded to us by the officers and their wives at these isolated Army Posts, I personally, of course, recall but little. Such things a small boy takes in his stride, blessed as I was, and undoubtedly with both of my brothers greatly spoiled; there is left no indelible brand upon his brain or whatever is the recording instrument where such things are noted down. But many, many times have I heard my parents speak of this unselfish happily offered succor to their jaded selves. The much greater security we enjoyed across the State of Texas we owed to our departure from the first Army Post, Fort Concho, being made to coincide with that of Major Wasson, the Army Paymaster. He was about to make his monthly trip when we got there, so that "the ghost might walk" for the doughboy and the cavalry. The Paymaster's escort, which accompanied us in consequence, was a safeguard against the Apache and other Indians, who at that time had not yet yielded to the blessings of civilization—or was it the Henry rifle?

I have a dim picture of some dashing officers, riding alongside the ambulances, to attend us for the first few miles beyond some of the Posts. In after years, I have realized that the true reason for this attention was not an interest in myself that I might revel in their handsome mounts, as I had then supposed; but that four very attractive sisters between the ages of thirteen and twenty years had more than a little bit to do with the consideration shown to three messy, normal, hence annoying small boys of the ages of eight, five, and three years.

As the journeying progressed the wretched pups had outgrown their crates, so these wriggling, squirming pests were then added to the discomfort of the grown-ups. One especially gallant young officer had given us constant and visible proof of his devotion to my third sister by insisting that we should take with us yet another puppy. This one was a strange animal of the Texi-cactu-mongrelous breed, whose life came to an untimely end in Chihuahua shortly after our arrival in that city. At this place of rest and enjoyment when the family might better have taken pleasure in his cute ways, this pup's life terminated.

The incident of his demise was the first demonstration of any homicidal tendencies in the Kid, his attempts heretofore upon my own life and that of his oldest brother being accepted as the regular tendencies of a boy of three responding to the life-giving rays of a Texas sun. We, therefore, were more than a little shocked when he dropped this poor unfortunate pup from a second-story window to see if he would land on his feet as the cat had always managed to do under like conditions.

It occurs to me here that this desire for scientific accuracy of knowledge on the part of Conness was the cause in after years of his likewise dropping a man who annoyed him out of a convenient window. . . .

II

Arrival

We must backtrack . . . we have arrived in Chihuahua with far too little travel and trouble.

The last days before we reached Presidio del Norte seemed to be the most trying and tiresome of all the pilgrimage. This was a small place on the banks of the Rio Grande. There was an adobe building or two, an American Consul, Mr. Russell, and some other kind people, among whom was Mr. Moses Kelley, who was extremely thoughtful and attentive. He did his very best for us on the United States side of the river where he had an adobe warehouse. Among his attentions was an immense tin bathtub, which was filled and emptied many times for our numerous baths in the muddy waters taken from that river which separates the sovereign State of Texas from the Republic of Mexico. At that period this river was a reminder more vivid than now of the amounts of powder exploded and the quantities of blood shed in the establishment of this boundary, where upon the day of our arrival the mercury stood at 115° in the shade!

The next morning to us was comparable in importance to the Fourth of July or any other great event in the history of the world. It was on this morning that we forded the Rio Grande, which was very wet.

The teams of six or eight mules to each wagon and to each ambulance were doubled in order to insure us against sticking in the mud and sand. Mr. Kelley led each division, driving in his especially made high-wheeled buggy drawn by two powerful mules, and the feat was accomplished. This eventful incident in the lives of my tribe brought to a successful conclusion, we reached the land of Mexico in which country more than one of the three generations of our family were to spend a good part of their lives in work and comfortable living. There we made friendships, some of which are still strong, but, unfortunately far too few, because many of those friends of childhood and manhood have passed to where beyond these voices there is peace.

When the requirements of the customs had been fulfilled, we resumed our trek toward the city of Chihuahua, the capital of the state of the same name whose plains and mountains, villages and cities make up the land which I know better than I know the land of my birth. While we are yet near the customs house and the town of Presidio Del Norte, it is fitting to point out that the innate and active courtesy of the Mexican race, its government, and its customs officials was then, as has always been my experience, something that could advantageously be imitated by those of the older Republic to the north.

As the crow flies, it is about a hundred and fifty miles from Presidio del Norte to Chihuahua, but we, alas, were not crows! Of necessity this latter portion of the trip was as slow and tortuous as the part which had gone before. But miles are only miles after all, and they can be worn off. One eventually reaches the point aimed at, or, if careless of the road signs, somewhere near the aiming point. Our bull's-eye for this first stretch of our Mexican journey was the city of Chihuahua, but before we gained that destination we came to another place which was but a short distance from there—Bachimba.

Ah, Bachimba! Bachimba!

What a paradise of irrigation, *caramba!* A garden . . . a dream of loveliness! It was a typical Hacienda, the first we had seen. It belonged then to the MacManus family. One of the MacManus girls had married the American Consul in Chihuahua, Mr. Scott. This gentleman had ridden his horse quite a way beyond Bachimba to meet us. Assisted by a cousin of Mrs. Scott, Ralph O'Neil, at the ranch proper he had prepared for the worn-out travelers a delightful reception in the cool, shady *patio*, a veritable oasis where fruits, vegetables, cooling drinks, fresh meat, eggs and chickens were in profuse and welcome abundance.

It was at beautiful Bachimba that we could once more delight in large trees for the first time in all these eight hundred miles of laborious travel. They were indeed a grateful sight to us who were accustomed to the gracious forests and parks of the East, to whom the dry, sandy, scrub-covered areas through which we had been passing had seemed not only useless but far from restful or prepossessing.

On the afternoon of June 19, 1880, we drove into Chihuahua. There we were housed in a comfortable large dwelling of two stories on the plaza opposite the stately old Cathedral. In her diary, my mother notes delightedly that there "was glass in the windows, a novelty."

In the city of Chihuahua we were the objects of every consideration and kind attention. After two weeks of settling us, my father and the other gentlemen of the party continued their journey into the southwest to Batopilas, where the silver mines were located. My mother and we children made our home in Chihuahua for a few months until my father could send back for us across the three hundred miles of plain and mountains which was the distance between the city of Chihuahua and the mining camp of Batopilas.

Taking a glance at my mother's diary once more, I see that on a certain day before my father left us in Chihuahua, they received a call from the priest, who came to pay a visit of courtesy, although they were of different religious faiths . . . quite different, for my parents were Presbyterians.

It was with no little—though carefully disguised—consternation that my mother witnessed the unrestrained and frank enjoyment and relish by El Padre of Dad's very excellent and choice Madeira. Wine at this time was brought into Chihuahua over miles of desert on pack-mules; it was an expensive luxury and difficult to obtain. There is no regulation in the Roman Catholic Church against wine, therefore it is no wonder that El Padre indulged his taste; but to my mother, accustomed as she was to the more reserved drinking by the good Presbyterian pastors of her former life, it was something of a shock.

At the price of seeming inconsistent here, I state that it was, however, entirely possible to experience a small jag from consuming two slices of mince pie made under my mother's directions from her grandmother's recipe. Never since my mother stopped keeping house have I tasted a mince pie "as is a mince pie," with plenty—and then a little more—of brandy and sherry. I suppose if brandy and sherry are cooked, their injurious and intoxicating qualities are eradicated!

Our wait in Chihuahua, I am informed by elders and betters, was an agreeable experience. Many pleasant persons were delighted to meet newcomers who were from a little known and a strange land. The difference in language naturally cramped conversational efforts, but there were one or two Mexican gentlemen who spoke English, and Mr. Scott of course. When calls were made and returned these patient men were requisitioned to go along or to come along as the case might be to act as interpreters.

The railroad was yet a long way from Chihuahua; it was not connected up at El Paso, Texas, with the Santa Fe until three years later on, and during those three years the only communication with their family and friends in the East which my mother and sisters could have was through letters which came once, and occasionally twice, a month. My mother records in her diary that during the few months when the men were in Batopilas, letter-writing was her chief diversion.

On the 16th of September, Uncle John arrived unexpectedly from Batopilas about dusk with the glad tidings that we were to pack up and come back to Batopilas with him, three hundred miles from Chihuahua.

Arrival

I can barely recall that there were Shetland ponies back home in Washington at Bleak House. I was now and then held on the back of one of them, when he, snail-like, ambled about the driveway confining the lawn. The past forty and more days of travel had fully impressed me with the fact that mules for progressing over the face of the earth were a great deal more accomplished critters than ponies if they were attached to an ambulance by a good strong harness, and stimulated by a whip in the hands of a "Dolf" or other skillful operator.

But now I was, with those others who for their great sins had the taking care of me, to appreciate the fact that the mule when ridden or better, sat upon, was a long-suffering safe means of advancing into the unknown. I differentiate purposely between sitting upon and riding an animal. There does, in truth, exist that dissimilarity: if you do not believe me, ask any Mexican saddle-mule.

On this my first journey into Batopilas upon the back of a mule, I shared the saddle, a small part of it, with a Mexican mozo, who was entrusted with my—to my parents at any rate—valuable carcass. I do not remember the name of this unlucky man. But I do recall that he had a long beard and that he kindly refrained from what must have been a burning desire to hurl me over the edge of one of the many tempting gorges with which that old trail is so bountifully supplied and along which it often skirts for miles.

If you are inclined to regard it as a pleasant or an easy pastime to carry a husky child in front of you while riding a mule over the Sierra Madre Mountains, only try it. Get someone to lend you a child about four or five years old. Procure a mule and a saddle. Go to the Grand Canyon of the Colorado and ride up and down the trail there for seven or eight hours, or until such time as your speedometer indicates that forty miles of Hell have come to an end. Do this for five consecutive days and then . . . well, I leave to your own best efforts the decision as to your attitude towards the child.

Of course, there is the side of the child to this story. But the child at this age is more or less helpless. He can struggle or cry out, but he has no responsibilities; his arms do not grow so infernally tired. He can lean back against the grown-up sufferer to his rear. . . .

There is not so much bother connected with camp-making, pitching of tents, getting of meals, and all the rest of it on the trail as one would suppose, if things are properly organized.

Into and out of deep gorges on corkscrew trails, along pleasant table lands thickly timbered, mostly with great pines, until the end of the fifth day's travel. On that day we set out before daylight. At the beginning of the third hour you come to the edge of the canyon down whose precipitous slopes you must go, some six thousand feet vertically, along winding trails. It takes you three and a half hours of riding before the level of the bed of the river is reached. Fifteen miles down this stream, the Batopilas River, is the haven you started out to attain at the Baltimore and Ohio Railroad Station in Washington, many eventful months past.

If we look backward and very decidedly upward, we can see the edge of the pine-covered bluff from which we departed some three hours previously. Between that spot and our present location on the banks of the river we have passed through pine timber, then oak, then scrub oak, next small semi-tropical growth, largely inhabited by thorns, and finally onto the region of cactus of varying sizes, shapes, and general pestiferousness.

From this place the trail follows the mountain side and accommodates itself and its level in accordance with the mountain side as it varies from steep to very steep and to extremely precipitous. At last we reach the junction of El Arroyo de Las Huertas. This is forded. On again for a short distance and at last we are come to the Hacienda San Miguel, which has been the true bull's-eye we have been aiming at all the time.

III

First Impressions

The Hacienda San Miguel was a perfect paradise for children. There was room and to spare. It was walled in against the high waters of the Batopilas River during the rainy season, for this river could rise twenty feet in a few hours. There was a big outside patio where grew immense old orange trees, oleanders, guayavas, *Noche Buenas*. . . . There was one huge *mata de aguacate* whose fruit we know as alligator pear. The beauty of crimson glory, the *Noche Buena*, is so called "Good Night" because this is the name for Christmas Eve throughout Latin-America. We know it as the poinsettia. Those in the patio of the Hacienda San Miguel were not merely plants, but were true trees with trunks of more than four inches in diameter.

We were all too tired the first late afternoon of our arrival to do anything but pay brief heed to the hearty and affectionate welcomes from my father and the others of the advance party, then to supper and bed. Ah! but the next morning! Then I set out to explore the hacienda with its irresistible mysteries. The noisy stamp mill and the whirring amalgamation plant, the machinery of which was driven from the hidden pit in which were the invisible turbines at the bottom of the waterfall from the aqueduct above. . . .

That small mill as the years passed grew into a big plant, and it was the first of three, but not the largest.

It was many years before the secrets were solved for me, the reason for those two revolving vertical pieces of shafting with their intricate cog-wheels and gears made plain. Childlike and manlike I hated to announce my ignorance by asking questions. They all appeared intensely interesting and correspondingly dangerous.

The last romance of Dora's life was mixed up with those pinions and gears. Three years after I first saw them, the boss of the amalgamation plant, Jack Ogden, carelessly allowed his right hand to be taken in, and it was so cruelly mangled that Dr. Ross was obliged to take it off. A stump was made for Jack and he used to poke me playfully in the ribs with it. He attached a useful hook with a clasp to this stump and at mealtimes he could manipulate his knife with dexterity. I was all of eight years old when I had my first experience with assisting the doctor to dress wounds. I can clearly see that old stump of Jack Ogden's now before it had entirely healed.

Naturally Jack became a great hero and was made much of by Dora who added many a good thing to eat to what Jack was supplied by the regular mechanics' mess. Very shortly after our arrival in Batopilas, Jack and the head carpenter began to pay assiduous attention to Dora. But the latter was only a runner-up. Dora married Jack and they moved eventually to Chihuahua where they took over the management of a hotel. It became the most prominent and the most popular one for many years in that city.

I suppose there would have appeared to be many things perilous to children at the Hacienda San Miguel. But evidently those of our family must have adhered to the instructions they received from those whose duty it was to look out for them and to guide them, for none of us was ever seriously injured.

In a very short time, Conness and I were turned over to a young Mexican boy whose function it was to "ride herd" on us. Alex was a bit too mature to need this attention. Dora supervised the kitchen, and Mary did the general housework with Mexican women to help her. Old Domingo Dominguez did the heavy work.

There is an expression in Latin-America, brought with them by the Spaniards from Spain, which I feel applies particularly to my two brothers and me: "*Dios es grande*," God is great. This expression is constantly employed in a place where the daily life of the individual brings him into more or less dangerous situations. They say that you are not killed nor hurt because "*Dios es grande*." Not only does this saying apply to our early life and the avoidance of serious accidents to ourselves, but to what we might have done to others.

Almost as soon as I was able to understand anything, the utter foolishness of pointing a gun at anyone was impressed upon me. Only one time did I violate this rule and then I got such a scare that never since have I been guilty of aiming a firearm at a person unless the need was real and I intended to fire it.

Old Domingo Dominguez' duties were many, but the one coming first in the day's activities was the bringing of fresh water for the *ollas*, which were kept in a window where whatever movement of the air was circulating would play upon its sides and hasten evaporation, thus cooling the water, for there was no ice. On a certain morning I was in my father's room looking longingly at the forbidden rack of rifles, shotguns, and other implements of offense and defense. I was about nine years old and it had been drilled into me that it was a wicked deed and the action of a tenderfoot to point a firearm at a person unless it was in actual self-defense. Domingo came into the room carrying the five-gallon *olla* filled with fresh water. He put it down in the deep window and straightened up. The temptation proved too much for my youthful self-control. I succumbed to the sinful desire. I grasped and took down from its rack what appeared to be a muzzle-loader, double-barreled shotgun. I saw there were no caps on what I took to be the nipples of the barrels, but which were unfortunately firing pins. I raised the gun and aimed it at the slowly straightening figure of Domingo. When I pulled the trigger I tumbled over backwards, almost stunned with fright as well as physical pain. I could scarcely bring myself to look at the spot which I was sure could only be designated as that formerly occupied by the late Domingo Dominguez.

How astonished and how thankful I was when the old fellow seized me, stood me on my feet and rushed out of the room to return almost immediately with a broom and a handful of mopping cloths! *Dios es grande*, and Domingo had jumped backward when I fired, through the door into Dad's dressing-room: the load of buckshot had taken great effect on the *olla* and its contents.

This affair could not remain a complete secret, but the heads of the family were away from the hacienda for some reason or other. I suppose it was because of this absence that I had got up my courage to take down that damned shotgun. I never have felt the same about one since. As a matter of fact, I have fired one only a few times in my life. It cost me many of my father's cigars to keep Cook quiet. He had dashed across the *patio* from the kitchen hearing the noise from the gun and Domingo's yell. I squared old Domingo and stood by him some fifteen years later, when he got into trouble and lost his job. No one ever knew it, but for months during those days I used to slip the old fellow ten or fifteen dollars every now and then, not because he was blackmailing me at that late date, but because of the common feeling that exists between two delinquents.

The only guard needed to keep us out of Dad's private office was Turk. He was a beautiful mastiff, quite gentle with us, but at the same time very positive in carrying out his master's orders. He would stay in the doorway himself to keep out any intruders. He did this effectually: in a mild but perfectly firm way he would punch his big old muzzle in our stomachs and push us back from the door. Turk had lost his handsome mate. She had to be shot when a drunken mechanic fatally injured her one night by giving her a brutal kick. The mechanic had never before been in such danger of his life. My father, the *Patron Grande*, was very kind to dumb animals: Turk and Countess had come all the way from Washington with us and they were his joy and delight. I have been creditably informed that what happened the night word was brought to him about Countess had the makings of a real tragedy. That is the correct expression, I believe, when it is applied to taking the life of a so-called fellow human being, which my father almost did.

It was necessary to keep us children out of my father's private office during his absence at the mines or elsewhere, because at the age of seven to ten we used to steal his best cigars to present them to some favorite of ours among the arrieros or even at times to smoke ourselves. Our special spot for these manly diversions was the roof. The flat roofs, the chimneys, and the accessibility of Dad's big, very good cigars were a challenge that must be accepted by young men like my kid brother and myself. When he was of the mature age of six and I eight, having sneaked a couple of these cigars into our pockets, we would upon occasions retire to the flat roof and selecting a place behind one of the chimneys where the thick foliage of the orange trees made it impossible for us to be seen from that side, we would light up. Then we could luxuriate in a delightful aura of dizziness, not enough to make the experience unpleasant, but just a mild drunk, enough to revel not only in the delicious fragrance of the first inch or so of a good cigar, but in that slight giddiness that went with it.

This all came to an end because we were seen by someone passing along the upper *patio*, who looking down upon us told it to someone else as a joke. Of course, this joke was passed on and on until it reached the ears of those that had the power and inclination to see that this pastime lost its charms. Despite this precocious start all three of my father's sons never became smokers until they were over six feet tall and the youngest one seventeen years of age.

The things we did to make the lives of those miserable informers unhappy kept us busy for quite a while. Thorns from any sort of cactus discreetly distributed between the sheets of an enemy's bed are no mean revenge. Poisonous insects such as alacrans, or better still a hideous tarantula, make any man regret his misdeeds if he realizes that they are a penalty he is paying for these transgressions. If he cannot possibly prove it on the culprits his exasperation is something greatly to be relished by the promoters.

Our diversions for the next few years were primarily an unflagging interest in goats, burros, stray dogs, most of them very hungry; whittling mules from cigar boxes and arranging pack-trains with them. We made long spears with which we chased cats assisted by the dogs and some young Mexican playmates as savage as ourselves. We caught mice in the orthodox local manner. This was done by using an empty spool and an empty tin cup. Into the center hole of the spool we stuffed a small piece of cheese; the cup we carefully balanced on the outer edge, the body of the spool extending back into and under the cup. When the mouse would creep under the cup to taste the cheese, he would jiggle the spool sufficiently backwards so that the cup would fall and thus make him a prisoner.

We were constantly seeing large numbers of raw hides being brought in for many uses about the business. What was more natural than that, in our efforts to reproduce in miniature the life and work of the adults around us, we should skin these captured mice (after killing them, I will say that), stake out the hides—we insisted on using that word, and salt them down to wait until they were in the proper condition to fold and pack in the same manner that we saw prepared the cow hides that came into the storehouse. This business of ours was done on the quiet and kept from the knowledge of the family as was the too violent pursuit of cats.

Of course, we had a grand time when the oranges and the guayavas were ripe. The orange trees were very old ones, their limbs were high above our heads. We would procure long fishing poles, tie a short stick near the small end in such a way that it would form a sort of hook. While one of us reached up with the pole and hooked the orange exactly where it joined the twig and by a sudden jerk detached it, the others stood about to catch the golden fruit as it fell. If these perfectly ripe oranges were allowed to strike the ground from those tall trees they would burst open.

Another thing which was attractive for us was out of bounds, and, therefore, prohibited. This was to sit on the edge of the ore cars while they were driven at a trot to and from San Miguel Mine a mile away from the Hacienda San Miguel. Since stones were sometimes rolled down the mountain side by goats or other forces of nature, very often the cars would hit a stone which had lodged on the track, jump the track and upset. The contents of the cars would be precipitated over the side of the road, from which place there was a nearly vertical fall to the river, a distance of over seventy feet.

This, therefore, was considered dangerous by our parents and it was also damned hard on the seats of our knee pants. . . .

The dwelling part of the Hacienda San Miguel was in the shape of an L. The bottom arm made the upstream or northern end and the vertical arm formed the side which would have been against the retaining wall of the mountain side were it not that, for the sake of light, ventilation and drainage, there was a ten-foot alley way at the back between the building and the retaining wall. This retaining wall extended up about thirty feet to another narrow bench, from which still another wall carried up to the last level or Upper Patio. On the last level was the narrow-gauge track over which the ore from San Miguel Mine was transported in small ore-cars drawn by mules from the mine to the bins of the fifty stamp mill. Over beyond these bins was another area upon which the thousands and thousands of mule-and-burro loads of wood were received and measured. When the wood was measured and the receipt for it given to the woodmen, it was thrown down a distance of seventy feet to the level of that part of the hacienda, the down river side, where were located the boilers, sawmill, lixiviation plant and roasting furnaces, as well as the terminus of the aqueduct which emptied into the turbine pits.

The dwelling houses of the hacienda were of one story, of solid rubble masonry, with flat roofs covered with mortar and brick. There was a wide portal in front and between the portal and the river wall there was a garden. It was a lovely, spacious place at least two hundred and fifty feet long and about seventy-five feet wide. A double row of orange trees and guayavas grew in the garden just beyond the edge of the portal.

IV

Murder

As I look back through the years there come crowding forward so very many recollections, so very many incidents that it is with perplexity a selection can be made best to illustrate the life of a young North American boy growing up in that entirely new environment. Yet that is not really true as we analyze it. What we now call new is not that to a little child; his customs and habits are not crystallized enough to be dignified by the definite remembrance of any one thing which would make the new particularly strange. The difference in language means nothing; the point where he begins to speak the foreign tongue is reached early and easily and so is not apparent to him. The acquisition of local habits and customs is about the same as was the learning of those that went before in the two or three conscious years antedating the change.

Until the time in 1883 when General Porfirio Diaz took command of the Government of Mexico, the so-called established government was a disorganized affair. Distances were very great between the capitals of the states; means of transportation were limited and very slow. Each small city or town maintained order as best it could with the restricted means at hand. There was no central power which had the capability to enforce the law in this isolated region when a band of *pronunciados*, of sufficient size to overawe the few local police or soldiers, appeared. During the first few years of our lives in Batopilas, every once in a while there would be an alarm of "*Los Pronunciados!*" Then all the men within the hacienda, American and Mexican employees, would go to the main office to receive arms and ammunition. There would be extra night watchmen placed at the proper places. The *Jefe Politico*, Don Jesus Hernandez, would ride up on his fine bay horse from the town. He and my father would agree on the methods of repulsing any attack that might be made, whether it be made from up river or down; in the former case our hacienda would be the first to feel it, if from down river then the town would feel it first.

Arrangements would be made for sending immediate messengers between the town of Batopilas and the Hacienda San Miguel, a mile apart. The canyon was an extremely narrow one, space was very scarce, and the road leading from the hacienda to the town was actually one long street. Wherever it was wide enough to permit of it, there was a dwelling or a store of the simplest character on either side of the road. It was generally above highwater mark unless there should happen to be an unusually heavy rainy season with a cloudburst up the river. As I said before, this was the only long street benched or cut into the mountain side.

Those *pronunciados*, as the name implies, were a group of men who were pronouncing against or rising against the constituted authority. They carried a flag of some color, often red or blue. The color meant nothing in particular, it was simply a symbol; and in this case red hadn't its present-day significance. They designated themselves as the "reds" or the "blues" or any other color that they had selected or found conveniently available in a piece of material large enough to serve as a banner under which, or better, behind which, to march.

However, I very much fear that the object behind the *pronunciados* was of no political significance, but was simply the desire to take away money or silver or supplies. They seldom were determined enough to put up much of a fight if the resistance amounted to anything; or if they received word that the town was armed and that we were prepared to give them a warm welcome, the threatened attack never materialized.

This was very exciting to kids from ten to five years old. It was the sure reason for the strapping on of wooden pistols, knives, the shouldering of wooden guns, which had the *pronunciados* only known it would probably have been quite as effective as the true weapons in the hands of some of the tenderfoot mechanics. But concerning this, as well as many other important and much-debated historical facts, the truth will never be known. Since General Diaz was now in full possession and Don Jesus Hernandez was resolved to fight to the last ditch, the *pronunciados* saw that an inflexible defense was certain and they gave it up as a bad job, returning to their homes in the mountains or in the Hot Country to the south and west of us.

We were about six, eight, and ten years old when the first tragedy came into our lives at Batopilas. Young Charles Mayhew, an exceptionally fine young American, was killed, shot in the back by a renegade Yaqui Indian who had been working at La Descubridora Mine where Charlie was timekeeper and assistant superintendent.

Mayhew was on his way to La Descubridora from the Hacienda San Miguel with several hundred dollars in his saddle-bags for the weekly payroll. La Descubridora was fifteen miles from the hacienda. You rode down river through the town, turned up an *arroyo* to the right, continued along this, climbing steadily for some miles until an altitude of some forty-five hundred feet was reached. Here the trail left the *arroyo* and made a laborious ascent to a small *mesa*, where the houses of the workmen were located with the mill. Across this another steep hillside climb brought you to the mouth of the mine close up under the bluff at an altitude of some five thousand feet.

Apparently when Charlie had been within two or three

miles of the mine, just before he left the old trail to the Hot Country, the Camino Real, which he had been traveling, he was accosted by one of the workmen and detained in conversation by him. When this took place, a big Indian named Alejandro, a one-eyed devil who had a reputation for violence and was wanted in Sonora (this developed later) for murder, stepped out from the bushes behind Charlie and shot him in the back without a word. As he fell from his horse, Charlie drew his gun and fired at the man facing him with whom he had been talking, realizing evidently that it was a hold-up. Charlie did his best as he dropped fatally wounded to retaliate upon the bandits.

This unexpected resistance so disrupted their plans that they fled, leaving the money untouched in Charlie's saddle-bags.

A woodman with his loaded burros on the way to town came upon the still warm body. He left his burros to the care of his small son and trotted on foot the ten or twelve miles to town to notify the authorities.

Immediately Don Jesus Hernandez sent word to my father, and himself followed closely on the messenger with several other mounted men. Quickly the posse formed and started out by two different trails in the attempt to cut off Alejandro and his companion. The number of these was not exactly known at the time. Only one person was believed to be with Alejandro, however, since all of the workmen except himself and one other were at the mine awaiting the arrival of the payroll.

A group of soldiers had been dispatched from the Jefatura as soon as word of the murder was received, with orders to follow the trail to the Hot Country. Next morning, the posses returned with jaded mounts after eighteen hours in the saddle. There was nothing to report but that the culprits had evidently departed from the trail to take to the ridges in a northerly direction.

It is impossible on horse or mule back to follow an individual who knows the country and who takes to the mountains, only a human being on foot—or a goat—can do this.

The excitement and interest aroused in our minds, the speculations as to how and when the pursuers would bring in the murderers, in a few hours, as is the case with most children, dimmed the sorrow for the untimely death of the young gentleman who had always been kind to us and for whom we felt nothing but respect. This sorrow, however, was freshened whenever we were in the presence of my father and mother. We knew instinctively that they were suffering, and that their feelings of responsibility to the parents of young Charles Mayhew and his family, and the letters they had to prepare for the next mail advising them of this casualty, were heavy on their minds and souls.

There was a lengthy consultation between Don Jesus Hernandez and *El Patron Grande.* A messenger was seen hurrying away from the main office, and shortly afterwards two individuals mounted on mules appeared with their blankets rolled and tied at the backs of the saddles. They were armed with .44 Winchester carbines in boots, and six-shooters. The names of these two riders were Andrez and Martin, both with reputations for being men of valor, "*muy hombres*"; they were about twenty-five years old at the time.

After a deliberation with Don Jesus and *El Patron Grande* in the main office, they emerged and, mounting, took the road through the main gate of the hacienda down the cobble-stoned ramp to the river. This they forded, and followed the main road leading up the canyon. My father offered a big reward for the assassins dead or alive. He and the *Jefe* believed that the two men sent after them would be successful in capturing them, particularly since the reward was considerably more than the amount of money a good miner would earn in a year—not less than a thousand dollars.

In discussing the matter that Saturday night on the plaza and in the cantinas, Andrez and Martin concluded that Alejandro was the instigator of the crime, and that his companion was El Mocho, so called because he lacked two fingers to his left hand. The two men had not shown up to draw their pay, they were not to be found at their homes, their women had not heard from them since they left on Friday night, when they had gone to work on the night shift. On Saturday when the miners came off shift, they would remain at the mines until the afternoon when the payoff was made.

Alejandro was called El Tuerto because he had but one eye. He was without money, so Andrez and Martin believed that being a Yaqui Indian he would want to go to a section of the country where he was known. This meant that they would travel north. When the murderers had covered what they considered to be a safe distance, they would return to the Camino Real, since on the trail alone was there opportunity to secure food.

Blood had been discovered on the trail a few feet beyond the place where Charlie had fallen and died, and this would indicate clearly that one of the murderers had been wounded.

The deputies rode hard and fast. By three o'clock in the morning on Monday they reached Teboreachic, where inquiry of the station keeper brought to light the information that he had *seen* no one, it had been a very dark night, but his dogs had awakened him. They were excited so very greatly that he had come out from his house to investigate. Two of the beasts stayed with him in the yard, the other one did not and he could hear this animal's bark, furiously violent, farther up the trail.

"And," continued the station keeper, "you know there is a short cut which the Tarahumare Indians use which leaves this *arroyo* ten miles above this place. This short cut leads to Eurique, from there it is not so far to the other *barranca* and the Yaqui Country."

Andrez listened to this reasoning and agreed.

"Have some breakfast with us and rest yourselves. When

daylight comes I myself will show you the spot where this trail departs from the *arroyo.* You could not find it for yourselves, even I could scarce do so in the darkness. You need food, also it will not hurt your mules to eat something. Let us delay *un rato* by which time we will be able to see clearly."

The station keeper was an elderly man, very intelligent. While they were sipping their coffee, he resumed:

"I know every trail and short cut in this part of the country. Now, if I had occasion to come cross country from where these assassins found themselves, if I desired to go where you believe they are going, I would have come straight up the *arroyo* exactly as did those unseen men last night. It is the most direct line. If you keep down this *arroyo* for seven leagues you come out into the Arroyo del Cerro Colorado above the big bluff."

"But do you believe . . ." began Martin.

"Wait a bit. If they came along the ridge of the Algarin which they must have done—there is no other way they could have made it—they would have come into the Arroyo del Cerro Colorado at that very point and at no other. The bluffs make it impossible to get off the ridge at any other place."

Came dawn. The three men reached the place in the *arroyo* where it branched from the Camino Real. They abandoned the road to climb to the *mesa* above, keeping up the creek for about two miles. Presently they turned into a nearly impassable tributary of the *arroyo.* Here their guide left them and returned to the station. Cautiously Andrez and Martin dismounted seeking the few spots where tracks could have been made. They came upon marks of what was apparently the passage of two men. They followed these nearly invisible traces for two miles when they disappeared. The deputies continued seeking, seeking, and were finally rewarded by finding tracks indicative that one man only had gone on from there. These marks were very fresh indeed.

The lay of the land permitted their remounting, which they did, and hurried their weary animals as fast as possible. Within a few miles they were halted by the sharp report of a pistol and the hum of a bullet close to Martin's head.

"*Por Dios y la Santa Virgen!*" exclaimed this one as they did a disappearance act. Each man slid from his mule and drew from the boot of the saddle his Winchester carbine. Quicker than it takes to tell it, the place from which the shot had been fired, where signs of the black powder smoke still showed in the air, was covered by a series of rapid shots. Some of them spattered against the boulder from behind which ambush Alejandro had fired one of his few remaining rounds. El Tuerto fired once more from this vantage point at the motion in the bushes to right and left, showing him that Andrez and Martin were crawling around to either side of him. Within a few minutes, they fired with sufficiently accurate aim to manifest to Alejandro that the game was up, and that they were indifferent as to whether they took him alive or dead.

The love of life was strong in this renegade Yaqui. He knew that there is always a chance for escape from human captors; he also knew that there is never any escape or even a chance for escape when one is dead. Firing his last two rounds into the air, he called out:

"*Me rindo! no tiran mas,* I surrender, do not fire again."

"Throw your gun over this way and throw it as far as you can so I can see it," was Andrez' reply. As the gun fell into the bushes all three men stood up, Alejandro's hands were tied behind his back and the noose of one of the *reatas* was slipped over his head.

Led thus by Martin and followed by Andrez they returned by the way they had come to the station of Teboreachic. There El Tuerto was fed. He wolfed his food ravenously, having had nothing, he explained, since leaving the Arroyo del Cerro Colorado the morning before. At that place he and

El Mocho had taken the food "*bien poco por lo cierto*, a very little sure enough," from a couple of young goat herds.

"Where is El Mocho now?"

"*Que sepa Dios*, God knows!" answered El Tuerto.

"You left him where? We know that he was wounded and that is why he is not with you. Now, just where did you leave him?"

"Near the Arroyo del Cerro Colorado," replied El Tuerto with no hesitation. "He could no longer travel. He told me that a charcoal burner who lived near by was a friend of his and would take care of him until his leg was well enough for him to go away."

Andrez look at the sun and remarked:

"When the sun is there," pointing to a spot in the heavens where if you had observed the hands of your watch they would have been at two o'clock, "we will leave for El Real. With the moonlight to aid us during the first hours of the night, we can reach there before ten o'clock. Sleep, Indian, you will have to move fast on the way back."

"Loosen my hands so that I can sleep."

"*Bien.* But we will tie them in front of you. We will also tie your feet," was Martin's response. This was done. The son of the station keeper was delegated to act as guard while they, the pursued and the pursuers, slept.

While these were sleeping, the station keeper went to the corral, saddled his sturdy pony and took the trail up the *arroyo*. He had learned of the easy capture from Andrez. When Andrez told him they first had thought there were the tracks of two persons, the old station keeper had nodded his head and said:

"There is a cave not far from the place where you tell me this happened. While you sleep I will ride up there to see what I can see."

This wise old man with a lifetime experience in finding strayed animals, and in living in the Sierras, soon was off his pony and examining with the care of a veteran the footprints described to him. He presently tied his pony and disappeared into the bushes, alongside the trail, which covered the hillside. He rapidly climbed to the place where an overhanging shelf of rock, not visible from the trail, made a natural cave extending back in the hillside some twenty-five feet, there it widened out leaving two shoulders on either side forming two niches.

Inside there was insufficient light to allow him an examination. Before he had left his station, old Julian had thrust through his *faja* a piece of fat pine. This he withdrew and split its end in several directions with his knife which he kept also thrust through his *faja*. From the pocket of his trousers he produced a flint and steel with a bit of punk. He lighted the fat pine, held it upside down until it was blazing well, then moved into the darkness of the right-hand niche. There could have been heard a sudden grunt: "Ugh, *asi lo creia*, as I thought!"

Julian spoke gently to the still body lying on the ground in the cave. There was no reply; he observed it closely by the flickering red light; with a shrug of his shoulders, he muttered:

"*Indio, esto es demasiado sangre fria*, Indian, this is too much cold blood."

For an instant he held the hand of the dead man in his own, it was quite cold, this mutilated hand. He peered yet more closely and turned the body over; this movement exposed the side fully . . . protruding from the ribs was a knife handle. He examined the legs of the corpse and discovered that one leg and foot were swollen and inflamed, which demonstrated that El Mocho had traveled miles on a terrifically painful foot and leg. It proved that Alejandro's companion had at last reached a point where it was utterly impossible to go any farther. They had found this shelter and had hidden there. El Tuerto had made up his mind to get rid of his friend who was now only an encumbrance. The simplest and best way to accomplish this was to stick a knife into his friend and thus to let the life out of him.

Julian took his neckerchief from about his own neck, unwound its folds and extracted his tobacco and *hoja.* These he placed in the crown of his hat. He shook out the neckerchief and put it quietly over the face of the dead man; he made the sign of the cross, left the cave, returned to his pony, mounted and started back to his station. Along the way he rolled a cigarette from his *macouchic* in a cornhusk *hoja.* Getting out his flint and steel as the pony ambled down the trail, he struck a spark which fell on the charred end of his bit of punk. This he applied to the bent end of his cornhusk cigarette, and when the tobacco was burning satisfactorily he obtained great comfort from its fumes . . . for fumes they are indeed. *Macouchic* gives off fumes, you do not have to smoke it in order to prove this, only ride behind a man who is smoking it! When Julian reached Teboreachic he went about his own affairs, not waking the tired deputies.

"Why should I waken them? There is no doubt of the identity of the man who lies dead in the cave. His left hand is mutilated as they described."

There was no need to waken them. The young guard was anxiously debating within himself as to whether or not the sun was in the exact spot in the sky indicated by Andrez, when this one came alive. He rose, wakened Martin with a touch of his foot. The other came to his feet with a startled murmur.

"*Vamos a comer y largarnos de aqui,* come let us eat and leave this place."

Speedily the meal was in readiness. Julian's wife called her son to take it from the kitchen; it was forthwith dispatched with gusto since it evidently was as savory to the palate as it was to the nose.

While Andrez and Martin were saddling in the corral, Julian made them acquainted with his finding of the murdered body of El Mocho with the knife projecting from the side.

"Well, he is better off dead now than to be shot later on," commented Andrez. "I will notify the judge. He can come out here to see the body for himself. I will not attempt to take it with us. I do not like that fat judge anyway, and this ride will make him thinner." They rejoined the Yaqui and his youthful guard.

"You, Yaqui *condenado*, be on your good behavior. On that alone depends your safe arrival alive at the Jefatura. For," he further elucidated with a shrug, "we will get the *mil pesos* of *El Patron Grande* whether you reach there alive or dead."

These comforting words were being addressed to El Tuerto while his feet were being untied.

"And," added Martin, "for the good of your soul you will travel these forty miles *pie a tierra*, foot to the ground. The *reata* around your neck will have its end on my saddle horn. Andrez will be behind with his *pistola* and his *carabina*. All of these things will induce you doubtless to have the proper care of your actions."

They made their good-byes and started off. El Tuerto was rested and strengthened for the trip by his two ample meals and four hours of sleep. He seemed to be perfectly willing to proceed, realizing that his chance for escape from his two captors was slim indeed.

Contrary to first instructions, they traveled slowly and rested before they reached the place in the canyon opposite to the dump at the San Miguel Mine. Neither Andrez nor Martin wished to miss the opportunity to demonstrate that "they had their man" by conducting him down that long street of the town at the hour when men were on the move to their work and the women were sweeping the trash from their houses into the street, to be collected and burned later in the day, when the *chimilcos* were open or just about to open—in other words about six o'clock in the morning. Then the night-watchman at San Miguel Mine would see them as they came down the trail across the river; the occupants of the scattered houses on the roadside would have the pleasure of seeing both culprit and captors. *Las viejas* and *las muchachas* would get their eyes full. This did not fail of its appeal to them both, particularly to Andrez who was as adept in making attacks on the hearts of *las muchachas* as he was in swinging a sledge, and in knowing *las pintas* . . . colors of the different veins of ore when certain signs appeared which were a sure indication that a *bonanza* was near at hand. It did not hurt his standing nor strengthen the possible resistance in the fair sex to have this additional demonstration of his courage and capabilities.

The tragedy of Charlie Mayhew's murder, the sudden taking off of a young gentleman who was loved by many and respected by all, threw a heavy cloud over the family. We children reacted to it according to our natures and ages . . . the bringing of Charlie's body to the hacienda, the lying in state for the necessary hours before it was carried to the Campo Santo located on a small, less sloping spot on the mountain side directly opposite to the hacienda, where a spur of the mountain lost its angle of sixty degrees from the horizontal. A tomb was built there overnight by our expert masons, and there rest the remains of this fine young American.

This part of the chapter that deals with the death of Charlie Mayhew came to an end with the incarceration and later with the execution of this degenerate representative of a splendid tribe of Indians.

The Yaquis of Sonora and the Mayos of Sinaloa made up a large portion of those loyal workmen to which the successful development of that camp at Batopilas may be chiefly credited. They were splendid in physique, they possessed unlimited endurance and were perfect Spartans in courage. These men were all fatalists, believing that until your time comes you cannot die, and that when it does come nothing that you can do will prevent it. They made the character of men most needed when there was dangerous work to do.

V

Buried Treasure

In Mexico formerly the big holiday of the entire year was Holy Week, *Semana Santa.* Three days' holiday for all, and a whole week for those who were more pious and also financially able to go without work for that length of time. It was a popular time for picnics, and the favorite place for these was Satevo.

Satevo was a charming location where the gorge widened out, giving a broader space in the canyon between the river and the red and yellow porphyry bluffs that towered above Satevo for many hundred feet and practically enclosed this small spot on two sides. Here was the very old and very beautiful church built by the Roman Catholic Fathers more than two hundred years before. This church offers an excellent example of the ability and determination of these great explorers, pioneers, and settlement-makers in the most inaccessible places.

Those of us who have lived and carried out constructive work in the Sierra Madres under the greatly improved conditions of later days, can really appreciate the achievements of those deeply consecrated pioneers. The church stands at Satevo today as a monument to those men whose knowledge of so vast a number of things made it possible, despite handicaps that would appal ninety per cent of the persons today who have the task of establishing a place of worship. It stands there in its majesty and dignity of architectural purity, stability, and beauty, as a memorial to those men and to the Christian Faith and belief in Jesus Christ. A belief so deep-rooted, so profound that the symbol which such a conviction inspired necessitated an erection of a building for all time, even without the ordinary care that the very best of brick construction requires, which is almost too little to mention.

The Fathers put up this church with the aid of one of the most ignorant, one of the shyest of all the Indian tribes in the Sierras, the Tarahumare. They are not rugged as far as manual labor is concerned, but they have tremendous endurance for traveling incredible distances.

They selected the clay, molded the bricks and burned them to a perfect hardness. They broke down the calspar from the true fissure silver-bearing veins in that vicinity, burned this and with the sand from the bed of the river they made a lime-and-sand mortar which is absolutely binding and durable. Wherever it is encountered now it is standing firmer and surer than it did ten years after it was built. There exist in the Dominican Republic among other ancient buildings the beautiful cathedral, the palace of Diego Colon, son of the great discoverer, Columbus and the remains of his great fortress built not later than 1515, which prove how practically indestructible is this type of lime-and-sand mortar.

This gem of a church at Satevo with its arched dome, its two-bell tower—one of them is marked with dates proving them to have been cast in Spain before Columbus discovered Espanola—always inspired me, even as a young boy, with a feeling that never could be put into words, intangible, abstruse, intense.

My first picnic at Satevo was in 1883. Even then the crypt beneath the High Altar had been broken into and the niches on either side, where the bodies of holy men had been entombed, were torn down and their remains strewn about the floor. This piece of desecration had been perpetrated by vandals who hoped to uncover gold or silver hidden there. The only treasure was the spring sunshine splintering its long arrows of gold on the arched dome.

You will doubtless recall that when it was decided to expel certain orders for real or fancied wrongs, the plans for the expulsion were laid with greatest secrecy. There was to have been a concerted action on a stipulated day in all parts of Mexico, in order that the confiscation might be a thorough one, that they might secure all the treasure amassed by the orders for the benefit of the Federal Government.

The leader of this movement was unduly optimistic in his belief that such a vital thrust could be kept a secret. The Fathers were forewarned and they succeeded in gathering together a large quantity of their treasure and conveying it out of the country. Nevertheless, they were forced to bury huge amounts of gold and silver in various forms. This was done because although the men who were leading the confiscation failed in complete surprise, the information came to the Fathers not very long before the zero hour. That which was sent away had to go shipped through channels unknown to all but a trusted few, and it had to be carefully disguised. All this took time, and when the fatal day arrived there was still some treasure in Mexico belonging to the orders, some unmoved and some in transit on the trails. This had to be hastily buried or otherwise disposed of, and for many, many years it was the cause of interminable search and excavations.

There is no doubt that many hiding places were found and rifled of their contents; but most of these discoveries were never publicly acknowledged, because the law of the country gives title to the Mexican Government of a part of all buried treasure discovered under penalty of complete confiscation if attempt should be made to avoid division in accordance with the law.

When I was a little boy I used to love to look at the shelves in the room on the right-hand side of the High Altar. This room had served as a library, upon them were still a few ancient tomes, one or two adorned by exquisitely executed illuminated work. The parchment of which they were made had been manufactured by the Fathers themselves. When I returned there a few years later, all these parchment books had disappeared. Whether they had been taken away and shipped to the Capitol in the City of Mexico I do not know. I do remember that my father had offered to bear the expenses for this transfer. They were of great historic interest, were written by hand in Latin, and contained records of the church at Satevo and its monastery.

The buildings which constituted the monastery had their connecting door with the church proper through the library, and it must have been a well-ordered and extensive edifice. Unfortunately the monastery was built of adobe. It was made with tough hay for a binder and was perfectly dried in the hot sun, and the buildings were thoroughly plastered with lime-and-sand mortar, but they were partially destroyed at the time of the expulsion, and the rains during the decades that followed did the rest. Thus that portion of the settlement where the Fathers lived and worked is ruined and the non-existent walls of a non-existent garden, where grew fruit trees and flowers, are only barely discernible. One must use his imagination to recreate this one of the many symbols of superhuman efforts on the part of consecrated, religious men who constructed in these wildernesses such monuments to their belief in God Almighty and His Son, Jesus Christ.

On one or two places along the trail where you have in your fancy accompanied me, and at several Tarahumare Indian villages at greater or less distance from the trail, we could trace remains of churches and small establishments where missions had been situated. The priest of these missions taught the Tarahumare how better to raise his corn. With seeds brought by himself from far-away Spain the priest planted the first apple and peach trees in the New World. Although I believe other fruits were planted which would thrive in that altitude and temperature, the apples and peaches were the only ones that remained. And these were very old trees and young seedlings.

Often in the lonely *arroyos,* where now there are no signs

of former habitations, these trees give to the Tarahumare Indian two articles of trade purchasable at the mining camps. When the fruit was ripe, he would load his own back or his burro or his pony with apples and peaches to sell. The peaches were quite small and certainly not prize fruit, and so with the apples, but they were eagerly seized upon by both native and foreign inhabitant.

From time to time I have tried to recall the name of the old gentleman who had the self-appointed task of caring for and showing visitors about the church at Satevo. He was a courteous little man, about sixty years old, I should say, in 1883. He was short, very lean, wore a moustache turning gray and he had thinning grizzled hair. He lived in a tiny house up the slope towards the bluff where there was a spring, which was his pride and comfort, because without it water would have to be brought from the river half a mile away. Like its owner, the house was neat and pleasing. Here with his plump little wife and an Indian woman and her son lived my old gentleman. His livelihood was provided by a herd of goats, some chickens, a few pigeons; a little garden was watered from the spring during the dry season, water which was carried by the son of the Indian woman in two five-gallon petroleum cans on the respective ends of a pole balanced on his shoulder.

The few sales were made to the inhabitants up and down the river, and few, in truth, they were. These people made occasional trips to Batopilas some twelve or fourteen miles up stream, which place afforded their only source of supplies.

His interest and care for the church were touching: its keys were his dearest treasure, and he received no remuneration for his services. When visitors came to see the church, which God knows was seldom enough, his satisfaction and zeal in officiating were pleasurable to witness. The story of his coming to this place is interesting. It was told me in later years by that other fine, loyal gentleman, Don Romulo Rocha, the last of a noble descent. Don Romulo was a trusted employee of the former owners of Batopilas from whom our company had purchased it, and he continued on with my father, contented with his lot, giving these two organizations not less than forty years of devoted service.

According to Don Romulo, my old gentleman at Satevo lived in his youth in the city of Mazatlan, a port on the Pacific coast, a pleasant place with which it was my good fortune to be familiar during my life in Mexico.

The family of Senor de Satevo was a wealthy one, and prominent, before and during the reign of the Emperor Maximilian, 1864-1867. His father was an Hidalgo of the old school. Shortly before the battles for independence began to rage, there had appeared in Mazatlan a certain Colonel B. in command of that Department of Mazatlan under the Imperial government. He was a dashing individual not without courage, contented with his personal attractions and the consequence of his position. This Colonel B. was a Frenchman, very intelligent and a clever politician. He felt convinced that there would soon be serious trouble. He liked Mexico, and enjoyed a small income aside from his pay and its not inconsiderable perquisites. He was positive that if the Mexicans made a resolute effort to throw off the yoke of Maximilian there would be a radical change in the form of government. He was determined to be on the winning side of any conflict if there could be any such possibility. If the forces for independence from the French were successful he could easily go over to those victorious forces, thus becoming a patriot of the country where he intended to make his home and one of whose fascinating native daughters he intended to make his wife.

Unfortunately the young lady whose charms had captured the fancy of the colonel was the sweetheart of our Senor de Satevo and had been so all of her short life.

The father of this lady was an Imperialist. He liked his

cognac and far too frequently expressed himself forcibly about the "upstart opposition" when he was in his cups. The colonel cautioned him about these opinions being too vigorously uttered, and made himself so very agreeable during those interviews, that when he made formal request for the hand of the fair young daughter, his suit was favorably looked upon.

Among the aristocratic families the hand of a daughter was disposed of entirely at the discretion of the father, the young lady had no choice whatsoever in the matter. Senor de Satevo remained a tormented onlooker until his sweetheart grew alarmed lest she be forced into a hasty marriage with the colonel. She conveyed her fears to our Senor de Satevo. He felt impelled to take a hand, and this he did promptly and effectively. Relays of good mounts were arranged for along the Camino Real. He took into his confidence some two or three of his intimate friends, whose *haciendas* extended a hundred or more miles in the interior of the country towards the north.

He eloped with his young sweetheart; some of his servants accompanied them, and the maid of his future wife. In the darkness of the night they rode at a gallop away from Mazatlan, and their disappearance was not discovered until the next morning, thus those hours had been wasted for the pursuers. When the chase began, the two lovers were seventy miles inland being married by a priest on one of the *haciendas* belonging to a friend. There was nothing now to do but to send word to his parents and his wife's father. Fast-riding men set forth to do these errands and met the pursuers as these hastened forward on their fruitless mission. They took over the task of notifying the respective fathers, pocketed the letters sent by de Satevo, and returned to Mazatlan.

The married lovers were now fugitives, they could never go back to Mazatlan. Colonel B. was the power there now. It was within his jurisdiction to do things which would have made it unpleasant and more than merely unpleasant for Senor de Satevo. Then, as now, a wealthy widow was a desirable acquisition. The father of de Satevo came under the eye of the colonel, and the old *Hidalgo* was persecuted to such an extent that he saw it would be unsafe for his son to remain within striking distance of Mazatlan. He sent word straightway to his son by a trusted messenger, conveying a fat purse, to leave for the north. The young couple journeyed to the foot of the mountains at Choix in the State of Sinaloa. There they became enthusiastic to visit "that great and very rich mining settlement at Batopilas from which the Marquis de Bustamente had taken such fabulous wealth." This place was across the border in the State of Chihuahua, a distance of seven hundred miles from Mazatlan.

They made a leisurely trip of fifteen days from the town of Choix to Batopilas. They lived there for a while, where de Satevo lost some of his fast-dwindling money attempting to mine silver. Word was received there, too, of the death of their respective parents. His own property and that of his wife had long ago been confiscated by the new government, the ill-fated Maximilian had been shot, poor Carlotta had gone back to Belgium, and the independents were in full swing.

De Satevo did not care for the resident priest at Batopilas, and he did care very much for the fathers of the order formerly supreme at Satevo. There he moved with his wife and there he rusted out.

Ah, but did he rust out? I used to think that he did when I used to watch him showing visitors about the church, finding it hard indeed to reconcile a dashing elopement with my thin little old man. But now, when experience of many years has had its way, I acknowledge that I am assailed somewhat heavily by doubts with a capital "D." I have worked a great deal harder than he ever did; likewise so have those with whom I have been associated. We produced a lot of so-called treasure from Mother Earth. At present I am nearing the years whose load he was carrying when I first saw him, and what am I doing now, but rusting out with barely enough to live on and nothing to leave behind me?

The sun has reached the edge of the bluff behind the church. We must saddle up and ride home. So, *adios* until another day, Satevo, when we hope once more to talk with your faithful old guardian and to refresh ourselves with thoughts of the days when the good Fathers were here; when this spot was a contented community, filled with growing things, fruits, flowers, people going to Mass, kneeling before your High Altar in prayers. . . .

I wonder if those of you living in this neighborhood are as free now as those who lived here then? Where do your women go to pray, where get consolation from the confessions of their sins? Was not the influence of the padre then as strong to prevent transgressions of the law as that of the local magistrate miles away? Somehow, although you have held on to your faith to a remarkable degree, I cannot but feel that you have lost a great deal that those who were in Satevo under the good Fathers rejoiced in, perhaps unconsciously.

In truth, there is something deeply esoteric in the symbol of the cross, or else the cross would mean nothing in these late days. In those Tarahumare villages remote from civilization, you frequently come across, on the very highest peak within vision, the silhouette of a wooden cross hewn from great pine trees, standing stately, aloof, held in place by veritable tons of rocks laboriously taken there by the Indians themselves on their bended backs and piled about its base to hold the symbol steady.

Only at rare intervals following the expulsion did the people see a priest. A very few times he could get there to perform baptisms, solemnize marriages, and collect his *peso fuerte* or his *fanega* of corn for each of these functions. The

Indians could have fled to the farther hills during the coming visit of the priest to avoid their meager contributions. Do not think that they were ignorant of his advent. They knew about his making his toilsome way from settlement to settlement several days in advance. But with no trouble at all the priest could collect his dues for performance of the rites. There are few cases, if indeed there are any, where an Indian has told of a worth-while mineral deposit whose location has been confided to him in secret; or has divulged the whereabouts of anything hidden during the exodus, information about which he had been forbidden to disclose. This knowledge, which they undoubtedly possess, has been handed down from generation to generation, and some mighty smart hombres have tried in the past by bribes, by presents, by argument, by threats, to get some of it, and they have failed miserably.

VI

El Patron Grande and Batopilas

When a man in the year 1880 took over a mining business in a locality such as Batopilas, the difficulties were many. There was no rail connection for approximately six hundred miles until 1883, when the Mexican Central got as far as El Paso, Texas. Then the distance between the mines and the railroad was reduced to three hundred or more miles. But those three hundred miles had to be traveled and freight had to be transported across them, for the first hundred and eighty-five miles on the backs of mules, and the last hundred and fifty miles in wagons or ox-carts.

After a few years' residence in Batopilas, we accepted the unavoidable mountain trip to Chihuahua as a necessary evil, a part of the essential game of getting the silver bullion from the mines to the market, and, in turn, securing the safe arrival into Batopilas of supplies.

Proper houses and corrals were established, with a resident station keeper, at the end of each day's march. Therefore, feed was available for the pack and saddle animals; chickens, eggs, turkeys, sometimes even milk, for the two-footed traveler. This journey was made with exact regularity and on perfect schedule every thirty days accompanied by the natural fatigue that comes of five days in the saddle, but no more. Conditions for making these monthly trips as comfortable as possible were taken care of by the company. A two-room house built of stone was constructed for the *conducta*, with a kitchen always ready for use by our competent cooks who accompanied the bullion trains. There was a portal all the way along the front of the house. In one room were two big chests or bins in which were kept the cooking utensils, cups and saucers, plates, knives and forks, spoons. . . . In each room there were folding canvas cots for the comfort of the employees who traveled for the company, and for any guests who were to visit us, to save the trouble and expense of the extra mules which would otherwise be needed to transport all such needed equipage. So much for the ease of the two-legged traveler. Let us look for the comfort of the patient, steady mule who carries this traveler.

If he is to make forty miles each day for five consecutive days and to remain in good condition, he must not be turned loose to graze and to search for and find his own food, thus to spend a greater part of the night in satisfying his justifiably big appetite acquired by seven hours of climbing up and down mountain trails with a mighty heavy load on his back. Besides, if he is allowed to be loose to graze there is always the chance that he might stray for many miles and be lost. The herder would be compelled to lose valuable hours in trying to find him, this would mean a late start, or a day lost on the trail. Some wary old mules are experts in hiding out so as to escape a hard job. It was, therefore, apparent early in the life of the company that regular stations with accommodations for both man and beast were an absolutely indispensable adjunct for the orderly running of the monthly bullion trains. They always took not less than fifty bars of silver bullion worth about sixty thousand dollars; more often they conveyed a hundred bars; on especial and unusual occasions they had as high as two hundred bars or even more.

Since each mule carries two bars of silver, you will readily understand that these trains are made up of from thirty to a hundred or more mules. Now, that is a lot of mules. The corrals and the barns for the storing of hay and corn and the housing of the mules made no small plant at the end of each day's march. It was highly important to have a resident station keeper. He supplied the corn and hay purchased by him from the Tarahumare Indians within a radius of fifty miles or so.

The hay was native hay and a price was fixed. In bad years it had to be brought in from a greater distance. The station keeper maintained his home here, he had his family, chickens, every now and then there would be a cow even, his turkeys, which all afforded a variety in the menu for the travelers. The arrival of the *conducta*, which was the bullion train, was a welcome event in the lives of the station keeper and his family.

The town of Batopilas itself presents an interesting picture of the early explorations made by the Spanish priests and *conquistadores*. In truth, they were tireless and intrepid men. History has no greater example of determination and courage to endure danger and hardships than that exhibited by them. They arrived in this mountain gorge eight hundred miles northwest of the City of Mexico more than two hundred and fifty years ago. They had no maps, it was an unknown wilderness of the most rugged mountainous country . . . a series of huge canyons between mountain crags from seven thousand to ten thousand feet in altitude. In the gorge two thousand feet above sea level, they established this town whose old masonry buildings were enduring and in a good state of preservation in 1914. Some of them in the center of the place were of two stories. Five thousand inhabitants, and a few more, lived there during our era.

My father's job as vice-president and general manager of the mines was not so much of a sinecure. There were five major mines—The San Miguel, Roncesvalles, Camuchin, Todos Santos, and La Descubridora. The needed hundreds of tons of freight, including foodstuffs and clothing, made a big problem to be solved. He had to maintain a close and careful supervision of the work at the mines, the transportation of ores to the Hacienda San Miguel, the treatment of these ores, the extraction of silver values, the melting and refining of the silver into bars of bullion, and their shipment each month. When you get your silver extracted from the ores and in shape to melt into bars, you might imagine that the work was about concluded, but here is what happened from that point:

The silver came from the ores of five different mines. Each had to be treated separately and exact account kept of each lot of ore and the silver returns from it. Each mine was credited with the silver it produced. Later when the silver was converted into dollars, the dollars were credited against the total expense of operating that particular mine.

Let us say that it is near the end of the month and the silver must be melted down and poured into bars of seventy-five pounds in weight. It must be accurately weighed, numbered and assayed; the number is stamped upon it, also the exact weight in kilos and decimals; two check assays must be made from chip or dip samples of each bar. Your shipping list often was for two hundred bars or better, this was in columns. No. of Bar...Weight of Bar...Assay of fineness of Bar... Value in dollars of Bar.

Added to my father's duties in supervising these constant jobs was the overseeing of the big building program which was going on. We were building a dam three miles up the river; an aqueduct three miles long; putting in a low-grade mill of a hundred stamps, of a thousand pounds each: Pelton wheels to drive the mill; compressors, an electric light plant, and the Lord only knows what else, including a hospital.

In a stamp mill of fifty stamps, if you use Fru Vanners for concentration you have two Fru Vanners for each battery of five stamps. In the early days, a Fru Vanner was the accepted concentrating table. There was a great deal of shafting and counter-shafting which carried pulleys of various sizes and drive belts. One of our early jobs was cleaning these many feet of different-sized shafting, all revolving at one speed or another, keeping them free from grease, slime and rust; filling grease and oil cups. We were paid for this work, but it was given us first to keep our minds occupied, and our young hands out of mischief, and second, to train us to run any part of the business in the future.

The ores of Batopilas, first discovered as I have said in 1632, are specifically native silver ores. Seventy-five to eighty per cent of all silver values contained are pure native silver, occurring in irregular masses, plates, and crystals imbedded in the vein matter. This is largely lime. When the rich pockets or *bonanzas* are encountered there is always a great amount of calspar. The native silver crystals, which in their turn have been covered by this crystallization which is quite transparent, are entirely imbedded in it. It is very beautiful.

The veins of the Batopilas native silver district are true fissure veins. That is, they are the result of some excessive pressure of the hard rock geological ages past, leaving big cracks in the rock from three feet wide to say twelve feet. These cracks have been filled in by minerals carried in solution from the lower levels in the rock up through these cracks. The deposition of minerals contained in solution upon either side or wall of the cracks continued until these were filled in entirely with solid matter. That is what is known as a true fissure vein.

The solution in that district contained lime, silver, sulphur, lead, zinc, iron, and small amounts of antimony and arsenic. All of these elements are, therefore, found in the ores generally in a crystallized state. A mineral solution, when it encounters the condition appropriate for it to become a solid, is formed into crystals which are deposited upon the first available solid matter present.

The outcrops of vein matter on the mountain side when discovered more than three hundred years ago by the Spaniards often showed the native silver quite visible to the naked eye. It was, therefore, a simple matter to extract and commercialize it. When my father took hold in 1880, the properties had been worked as deep as they then could be worked by the primitive methods employed. On the backs of men it is possible only

Batopilas Mining Company's outfit (*La Conducta*) ready to start. Chihuahua.

Town of Batopilas, Chihuahua, Mexico looking up Batopilas River Valley, toward Hacienda San Miguel, on right, 1885 (about).

Main Street of Batopilas, with church at left.

Street between Hacienda San Miguel and Batopilas. Tower in background is the main office of the Batopilas Mining Company.

to carry ores up ladders for five hundred feet, more or less. The best peon, a laboring man, will carry one hundred to one hundred and fifty pounds in a *zurron* up a distance of five hundred to a thousand feet of ladders . . . but he cannot make many such trips in a day. Hence, the cost becomes prohibitive. There is always a greater amount of valueless material produced in mining pay dirt at Batopilas than there is valuable ore.

We used horse whims, then steam hoists; afterwards came gasoline hoists, and compressed air drills, and for many years the longest mining tunnel in Mexico, the Porfirio Diaz Tunnel, with shorter tunnels at lower levels on the mountain side, were the methods which were employed for over thirty-five consecutive years of activity in our camp. Our total mileage of underground workings was over seventy miles.

Low grade ores cannot be worked profitably unless the cost of production is reduced to its lowest point. The cost of working a mine increases as its depth increases, and the ores and waste must be hauled to the surface from great depths. The means to obtain economical power for treating low grade ores, the building of the hundred-stamp mill (each stamp of a thousand pounds), and the way to convey the ores to the mill were another demonstrated picture of my father's always looking ahead years farther than does the average man. The result was the Porfirio Diaz Tunnel. Within a hundred yards of its mouth was the mill and also was located the end of the aqueduct, whose waters after a fall through flumes for a distance of sixty-five feet and by the assistance of Pelton wheels, delivered some eight hundred horse power. The Porfirio Diaz Tunnel required several years to drive. It cut all that group of veins found on the western side of the river. From the Portal to the point where it tapped the Roncesvalles vein is seven thousand and fifty feet. It is continued on this vein and the Todos Santos vein for two thousand feet more. The old shafts and workings from above are connected with the tunnel; therefore, instead of being compelled to hoist them, the ores roll down to tunnel level, making a much more economical proceeding altogether. From this place they went direct to the bins of the mill.

The rock of this country is a diabase and very hard, far from easy to drill but it breaks fairly well.

A visitor from the north described the life at Batopilas as a perfect type of the old feudal days. I know that there was a perfect discipline founded upon respect and affection for those in command; an intelligent set of rules impartially enforced; good pay and a personal interest taken by *El Patron Grande* and the *Patroncitos* in the lives and welfare of all the employees.

My father perceived almost from the beginning that it would be necessary to found a hospital and dispensary, not only for the American and other foreign employees but for the natives as well. It was, however, with difficulty that they could be persuaded to attend; this complication of affairs lasted for some months after things were established. We made no effort to force them, and gradually they began coming of their own accord to be helped in minor illnesses and to be carefully nursed in major ones. Any man badly injured in a mine accident (such cases were comparatively few) was taken at once to the hospital, willy-nilly, and kept there until he was completely cured.

Dr. Ross came down with us from Washington, and was our first medico, a good man. There came next Dr. Lee, followed by Dr. Frank Merchant, our first cousin, who married my sister Grace. Dr. Charles Morrow came next who afterwards became a surgeon in the United States Army. I ran into him at Robert de Espagne in France in 1918, he was Divisional Surgeon. Dr. Robert Wagner came after him, and he married my sister Isabel. Phil Riley was the last of our medicos, and he now wears the Army uniform of Uncle Sam.

The complete hospital was built about the time Dr. Robert Wagner came to Batopilas; two able professional nurses came from the States to take charge. A clinic was instituted and both employees and non-employees received skilled medical care and attention free of charge. This, combined with plenty of good food and other care, brought the health standard of the camp to a very desirable state. The men who functioned during those thirty or more years gave an unexcelled proof of what a competent and conscientious M.D. can do to benefit humanity.

El Señor Doctor was a person of great and deserved importance in the community, a strong influence for the general good. All of these men reached enviable positions of prominence during their lives and medical service to humanity in other places when we shut down in Batopilas. All of them with the exception of Phil Riley have passed on. They were comparatively young men when death overtook them, it is quite probable that the hard work, the confining duties of their constant jobs helped to wear out the machines that otherwise would still be running.

VII

Everyday Matters

If boys grow up in a country where it is customary to buckle on one's six-shooter as a matter of course, whenever one leaves the house, it is only to be expected that their greatest longing in life is to do likewise. Carrying your gun in a proper holster makes pulling it out smooth and easy with no risk of loss of time in the draw. It insures the gun's remaining at your side and not thumping Hell out of you, or flying out from an inadequate holster when you are riding your horse or mule at a dead run over a rough country where the animal, frequently in full stride, must make a cat-like spring up or down, or to one side or the other.

Those persons whose not-to-be-envied obligation it is to direct and control your actions during this adolescent period have a bad time of it explaining that a boy of six or eight years of age is entirely too young to "tote a gun," particularly a Colt .44 or .45. The yearning, however, is just as strong in the boy after he has listened to the argument against it as it was before he listened, in concealed boredom, to what the oldster considered a reasonable and able effort in debate.

A gun at your hip in a holster is an absolute necessity. You put in many hours whittling from a piece of board the outline of a grand old Colt .44 or .45, with a six-inch barrel, that has been obtained as a model by stealth while some lucky grown-up—one of the young uncles probably—was absent from his room and the cherished gun hanging in its holster on the bedpost.

But there is a substitute for a six-gun when the real article is beyond your attainment. This was very much to be coveted because its possession was secured by exercising acuteness in scouting and caused immense wrath among the carpenters . . . this substitute was a claw hammer. The habit of making use of claw hammers in lieu of six-guns was put down by my father after we had made a successful raid on a certain tool-chest in the storeroom during the owner's absence on his annual vacation.

Mr. Kemper was very tall, very angular as to physique, disposition and temper. He valued his tools to the point of an obsession; they were as fine and complete a set as any master carpenter could possess and they justified his pride in them.

We filed away two massive padlocks. We lifted the lid, removed the well-filled drawers from his tool-chest, and uncovered a veritable treasure trove that caused the three of us to swell with joy. This was all the more intense because we knew that Hell would pop as seldom it had popped before in our lives when the old boy came back from his vacation and realized that his precious chest had been violated. This return happened sooner than we expected and before we had rejoiced in even a short two weeks of toting two old and one beautifully new, nickel-plated claw hammers.

We will not delve too deeply into particulars. We succeeded in getting the hammers back into the tool-chest and kept mum. But when old Kemper, or Temper as we called him, saw his tool-chest, beheld the two massive padlocks filed off, witnessed the further defilement, he got red hot from his feet to the top of his head. In this state he arrived in the main office shivering with ire.

He really did forget himself and that my father was six feet two inches tall and weighed two hundred and thirty pounds of the most forceful and emphatic material he ever saw. Needless to say what took place between them was not seen by us, but we got a detailed statement from the bookkeeper, who hated Kemper, as did nearly every one of the other employees.

By the process of deduction we appreciated the enormity

of our crime when we were finally discovered by old Domingo. He was general errand boy and chambermaid for the rooms of the American employees. Domingo was fond of us boys and he gave us an outline of battle conditions. There was to be no escape for us. We tried to bribe Domingo to report that he had been unable to locate our hiding place, but his respect for *El Patron Grande* was such that this could not be accomplished.

Should we deny any knowledge of the facts?

"Impossible," stated Domingo. "The storekeeper in order to protect himself against suspicion of allowing you to take the hammers was obliged to tell your father that he had seen you boys with them. And," continued Domingo, "he is very angry with you still for taking his cats for your dogs to chase."

We went to the tribunal. Let us draw the curtain of charity over the rest of the episode except to state that we were the recipients a short time afterwards of three of the very nicest and most realistic, life-sized wooden six-shooters that could be found anywhere. They were presented to us by one of the under carpenters whose detestation of Kemper was so great he felt it must be manifested to his own inner satisfaction by working overtime to make these quite wonderful wooden guns for us. Poor Kemper was employed by the company for many years, but I honestly believe he always hated us with a deep and fervid hate, even when we were almost grown. He was, however, a very fine carpenter.

Several years after this a young American came down from the States to work for my father. Louis M. became enamoured of Kemper's daughter, a young girl unhappily endowed with a large nose, but with a figure that might almost be taken as a miracle when you considered the physical architecture of her father and mother. Louis' suit did not prosper, and the disgruntled suitor composed some verses for "Ta-ra-ra boom de aye," which jingle having made a hit in the United States came to Batopilas by way of Louis M. During the interlude between the claw hammer incident and growing older, I had been unable to conceive of any way to revenge myself on old Kemper. Therefore, it was with a feeling of complacency I used to sing one verse that went like this:

"I got a girl in Mexico,
Insect bit her on the big toe,
Now she's where the cactus grow.
The name of that insect you all know . . .
Ta-ra-ra-rantula!"

Louis M. was a talented young man some years older than ourselves, but he was very companionable and exceedingly entertaining.

About this time the Governor of the State was coming to make an official visit to this ancient settlement of Batopilas, of such great importance to the State financially and otherwise. Naturally it was to be a very particular event and everyone wanted to make the most of it. The records showed that never prior to this time had a governor made that arduous ride across the mountains to this place which had been famous for many years as a producer of silver. Great preparations were set on foot by the townspeople collectively and individually. Arches were erected bearing "*Bienvenida*" in big letters, others with "*Viva, Señor Gubernador.*" There were to be a series of *bailes*, *serenatas* with all manner of other entertainments; also speeches early, often and many.

The Governor arrived with a large staff of secretaries, a detachment of cavalry, one of infantry, his military staff and his servants. For nearly two weeks official calls, inspections, dances, *serenatas*, banquets, kept everybody busy. I was at the Prep school age and it is all as clear as yesterday. One special happening stands out in which Louis shone with his customary amusing light.

My father gave an official dinner for His Excellency. It was quite an affair. As usual there was the customary ample supply of all kinds of wines and liquors. The band was engaged to play, the moon shone softly bright, the orange blossoms perfumed the atmosphere with their fragrance . . . the setting was one that inclines the spirit towards liberality of thought in relation to liquid refreshment. The time came for the guests to take their departure. Elaborate farewells were being spoken and replied to by the host, my father. Louis M. was dispatched to the room set aside for the gentlemen's dressing-room, some distance down the long portal, to bring the Governor's hat and cloak to him. The party congregated at the steps where the mounts were assembled, held by *mozos* arrayed in their resplendent best. Louis did not appear, neither did the hat and cloak. Everyone waited. The Governor and my father turned to stroll towards the dressing-room engaged in close confidential conversation. They arrived opposite to the door, and Louis waveringly arose from the chair upon which he had surrendered to the effects of orange blossom perfume and other fumes. With the high silk hat of the Governor reposing on his own head he exclaimed:

"I am very sorry, sir. I do not think that His Excellency the Governor wore a hat, or else someone has taken it. I have searched for it everywhere."

In deep mortification my father seized the hat from Louis' head, apologizing profusely to the Governor. But this one was a good sport, and although he understood no English, he was entirely conversant with many other things . . . and their potent effects.

When I was still at Prep school age I used to ride, as did my brothers, to the different mines on Saturdays and sometimes on other days, to come down as guard with the pack trains of first-class ores. Then I was permitted to strap on a very small .22 revolver handed me by my brother Alex, who by that time was fourteen years of age and sufficiently responsible to be allowed to tote a .38 or even a .41.

On a certain morning I reported at the main office for the

payroll money. The expression of my father's face, quizzical, kindly, amused, touched, will long remain in my memory as he looked at the old .44 frontier Colt I had strapped on me. Although I was a very well-grown boy it nearly reached to my knee. I had inveigled the cashier in the front office to let me slip it from the arms closets which were under his care. After many hours of rubbing it and soaking it in oil, it was in a condition to kill anyone or anything as effectively as any newer and flashier gun. A big part of my savings for months had been invested in the box of fifty cartridges . . . they cost five dollars for fifty. When I got that old loop belt cut down to fit my waist, had filled it with .44 cartridges, and was seated on my pony or mule, my pride was such that I would have taken, I actually believe, a chance at firing that gun even if attacked by Billy the Kid.

We used to love to play *palio*, it was a favorite game with us. *Palio* is a vigorous form of the game of hockey as it is played in Ireland. The ball is a solid wooden one or it may even be a round rock procured from the conveniently close river bed. The stick is a shorter one than is the regular hockey stick, heavier, and it is wider too; instead of a hook, the end is slightly bent with a spoon-like effect. It is thick and weighty enough to scoop the ball into the air, and before it hits the earth you can drive it farther still by belting it lustily. It sometimes happens that the ball may be sent back by an opponent before it hits the ground. Quite a large number play on each side, the grounds are usually alongside the river and are covered with all sorts of stones and gravel and coarse sand. It is rough going since you generally play in a shirt and pants, wearing an ordinary cap or hat, or maybe you are bareheaded, with no guards. It is decidedly not a game for a mother's darling or for boys or men who allow themselves to feel pain unduly. It is not uncommon to be knocked flat and stunned by a blow on the head. The young North American soon learns to be as Spartan as are the young Mexican and Indian.

To be made to groan or to cry out is to be branded by a certain name that is regarded the very last thing in the way of high-class insult. I have often seen a fallen warrior, thus offended, rise from the ground in a fury of rage only to beat and bang the offender into a recumbent position in his own turn. The first individual is thus reinstated to good standing. It is harsh, boisterous; but it is very good training. Some of the cracks on the head may be the cause of certain eccentric actions you may be guilty of performing in after life, but I do not really believe that such is the case . . . those delinquencies would better be laid to normal general cussedness.

When your early youth is surrounded by swearing in two or more languages—there were Italian stone masons, German mechanics, and others on the payrolls during those first years—you acquire a vocabulary filled with terms unprintable although appropriate to the many lines of endeavor with which you come into contact.

For instance, the *arrieros* have a poetic and charmingly vulgar line, but never do they take the name of the Lord in vain. The miner is almost as accomplished as the *arriero*. Then, of course, the stone mason of Spanish or Italian lineage has his specialties. The inartistic mechanic, German or American, whose "cussing" efforts are not in the same class, has a much less esthetic effect. Estheticism played but a minor part in our lives.

My mother and older sisters tried to teach us three youngsters, and they did their best to make the lessons interesting to us, but we found the outdoor activities much more exciting. We watched the building of stone masonry walls or adobe plastering, the mixing of sand and lime for mortar, the blending of the mud-clay and straw for adobe. You could not drag us away when the workmen were riveting boiler plates or placing and expanding tubes in the boilers, or keying cams on cam shafts for the batteries in the stamp mills, or keying bossheads to battery stems, or wedging the shoes of the bossheads.

Everyday Matters

By the time most boys are considered old enough to play ball on a vacant lot we had acquired a general knowledge of material things and it was a very simple matter for us to take our places in the picture and assume the at first small jobs that *El Patron Grande* allotted to us.

VIII

All Kinds of Education

About the year 1885, the very nice, kind, long-suffering Miss Griswold came to Batopilas from the States as a governess for my two youngest sisters and for us three boys. Due to her efforts and those of my sisters who read aloud to me a great deal, I developed a fondness for reading at rather an early age. *St. Nicholas*, of course, came regularly to the Camp and we had bound editions of it for the years preceding those when I first learned to read haltingly for myself. Then I had boys' books, *Frank on the Prairie*, and *Frank on the Mississippi*, until my wise mother decided that such fiction might be supplemented to advantage by more instructive reading which would also inspire patriotism. One Christmas when the steamer trunks of Christmas presents arrived from New York, there appeared *The Boys of '76*, *The Sea Stories of 1812*, and many other thrilling accounts of historical incidents. There were *Young People's Encyclopedia*, one volume of *Persons and Places*, and one of *Common Things*.

Miss Griswold exhibited great determination and courage in standing the strain of teaching us for some time, and we were forced to waste on the acquiring of preliminary book-learning many hours which could have been devoted to *palio*, mules, horses and other exciting things. But after two years of us, poor Miss Griswold gave us up for lost and returned to the States. Our education then continued at home for a while.

All the family, except my father, returned to Washington in 1886 for the double wedding of my two oldest sisters, which took place at St. Paul's Church, Rock Creek Parish. May married E. A. Quintard, the son of the Bishop, and Sue married W. M. Brodie, a Scotch engineer. It seemed to me a very large affair, I was duly impressed by the Bishop, and Bleak House was crowded. I remember as though it took place only last month the three-mile drive from Bleak House to the church; down the driveway, through the gate into Seventh Street Pike, the turn to the right past the Lay Place, where is now located Walter Reed Hospital, past Fort Stevens to Brightwood. In those far-off days there were an inn and a race track at Brightwood. We made a left turn into the country road known as Shepherd's Road, on past the farm that once was owned by the grandfather of the brides, and finally into the beautiful wide churchyard.

I recollect nothing of the drive back to Bleak House after the wedding, there I was completely submerged by the ocean of good things to eat awaiting us. We ate in the front of the house with the rest of the guests, and then we wheedled rich and indigestible foods from the servants in the kitchen and pantry. That splendid debauch we regretted some hours afterwards with all of our digestive apparatuses.

About a year later on, when I was twelve years old, another return was made to the States. We did not stay at Bleak House then, because it was winter time, but rented a small house. My mother made up her mind that the Kid and I ought to go to school and she selected the Friends' School. I hated it. For some reason I was a victim of shyness, and fear of the ridicule that might come to me if I failed when called on in class made every day a nightmare. The worst ordeal was spelling class. We were lined up, standing, and if you failed to spell a word you were sent to the bottom of the line. Such was the agony of this torture that I pretended to be sick morning after morning. Some mornings, I went to the School a moment too late, then I would come on back home telling my mother that the doors were closed and locked. Several times, my amiable sister enticed me into the pony-cart and drove me to the School where she left me, hoping that I would not run away. But usually I would arrive home before she did.

I ran away one morning, determined never to return either to the School or home. I wandered east as far as Fourteenth Street, turned north and went up Fourteenth for many blocks. I came to a store window in which were displayed some bantam chickens. I gazed and gazed until I grew so fascinated by the idea of possessing a pair of bantams that I made up my mind to strike a bargain with my mother. I would overcome the horror of the Friends' School if she would give me a pair of bantams. I do not recall what excuse I gave her for skipping school or what lies I told, but I did not get the bantams.

Needless to say, the Kid and I were delighted to get back to Mexico.

I think it was in 1887 that my father made one of the two trips away from Batopilas in the twenty-two years he was in that camp. Some of the family accompanied my parents to the City of Mexico, and the Kid and I went too. They spent two or three weeks at the Hotel Iturbidi, which had been converted into a hotel from the old palace of the Emperor Iturbidi, in order to see President Diaz, and to attend personally to a press of affairs in connection with the company's business. This was a memorable trip for the Kid and me. John Worden had changed his run on the railroad, and he was now pulling the passenger train from Chihuahua south for a long run, a full day at least. He had given our company good and faithful service before he was employed as an engine driver on the Mexican Central Railroad, and he was very fond of the Governor and of us boys. Consequently he permitted us to ride on the engine all day. That was a grand sensation, and did we simply fly! "I betcher" we made, in a few of the places where the tracks were good, as much as thirty-five miles an hour.

It was our first visit to the capital and we surely had a full and satisfying experience. It was there I had my introduction to the old-fashioned hat tub. On our first morning a couple of *mozos* came up to our room, the Kid's and mine, bringing a large round tub made in the shape of a wide-brimmed beach hat upside down. On one side of the wide brim were fastened two little legs, and you were supposed to sit on the seat-like arrangement, supported by these legs, place your feet in the crown of the hat, so to speak, and with a sponge proceed to wash yourself. They put the tub in the middle of the floor, poured five gallons of warm water into the crown and took their departure.

"This is a funny-looking bathtub," thought I. "But here goes!" I stepped over the wide brim into the center of the crown and sat down on the brim. But being stupid I had failed to realize that the place to sit was on the spot supported by the two little legs. Of course, I went over backwards accompanied by the tub and the water, and the squeals of delight from the Kid.

The next year when I was thirteen I was sent to the States to Lawrenceville, New Jersey, where Alex was now entering his second year. The masters at this estimable seat of learning, having had many years' experience with youth, had found it expedient to institute a preparatory grade for the first form. This was known as the shell. My stupidity as to lessons insured my joining some fifteen other boys as occupants of the shell. We enjoyed ourselves, had a pretty good football team, and since I was a big, clumsy boy, I made an excellent center rush. We played the second first-form team and often defeated them much to their mortification and to our delight. I was mighty homesick for Mexico for a while, but after that wore off, I had a good time at Lawrenceville.

There are many places I want to see once more before I shove off for the last cruise. I want to go back to Lawrenceville, see the early morning of an early summer day, the morning of the day when Princeton played baseball against Yale at Princeton.

I am anxious to compare the kick I will get from going over

to Princeton from Lawrenceville to witness the game at the age of sixty-odd, with the kick I got that morning of the eighth of May in the year of our Lord 1889. Figure it out for yourself. A kid between the ages of thirteen and fourteen, who, had he devoted the same amount of time to his books that he did to football and baseball, would have been in the first form instead of being in the shell. He was to go in a big 'bus drawn by none too fiery steeds from the Hamil House, which was his Hall, to Princeton to be present at a baseball game in which the batteries for Princeton were Snake Ames and Fred Brokaw; for Yale, Alonzo Stagg and Poole. What a day! What a day!

It was a hot day, it was a day when Princeton was victorious fourteen to eleven. My hazy recollection of that game and the players has been clarified by a letter that Asa S. Bushnell, graduate manager (Princeton), was kind enough to write me in reply to an inquiry of mine as to the make-up of the teams. Here it is, probably some of you may also be interested in the archeology of baseball.

	Yale		*Princeton*
p	Stagg	p	Ames
c	Poole	c	Brokaw
1b	McBride	1b	Dana
2b	Calhoun	2b	Osburn
ss	Graves	ss	Knickerbocker
3b	Noyes	3b	Watts
lf	N. McClintock	lf	Payne
cf	Dalzell	cf	Durrell
rf	G. McClintock and McClung	rf	King

I can see the field, I can renew somewhat the thrill as the game was called and things started. No two more different men than Snake Ames and Stagg ever faced one another as pitchers in any game. They were dissimilar in physique and in temperament; it was a hot game; there were times when

Ames grew irascible, after which he would wind up and his intensity of delivery would make you wonder why his arm was not torn from the socket and did not follow the ball to the plate. Stagg, more composed, sweated profusely as he calmly stuck to his knitting. Seems to me I see him in a knitted blue quarter-sleeved jersey with the Y on his chest. . . .

Lawrenceville sent many leading athletes to both Princeton and Yale. When I was there I can recall Pop Warren of football fame, Mitt Lilly, the Jacksons (brothers), Huss, Sam, and others who earned national reputations.

The term ended, and after a summer at Bleak House, my parents resolved to try a tutor for us. In the fall of 1889 it became the misfortune of Mr. Clarence Stoner to labor with the brains of the three Shepherd male children.

It was good to return to Mexico. Fourteen years seemed to me to be a dignified and mature condition of manhood, and our earlier methods of play were too childish for me. I had acquired a strong liking for reading, and spent as much time as I could with *Charles O'Malley*, the *Irish Dragoon*, *Tom Burke of Ours*, and all the rest of Lever. It was hard for me to decide whether Charles O'Malley or Tom Burke was the world's greatest hero. I was a strong admirer of Napoleon Bonaparte, but I could never work up an excessive rise in temperature over Wellington.

We had no idle time on our hands, which was undoubtedly fortunate, for the Lord knows we got into mischief enough as it was.

As I said before, on Saturdays and sometimes on other days, we were employed as guards for the pack trains bringing down the rich first-class ores from La Descubridora, Roncesvalles, Camuchin and San Miguel. Often we acted as guards and watchmen at the twenty-five-stamp mill while these rich ores were being broken up by sledge and hammer into fragments sufficiently small to permit their entrance into the batteries. After crushing in the batteries, the heavy silver which would not pass through a forty-mesh screen was cleaned out of the mortar bed. This required careful supervision for there were always heavy chunks of pure native silver and nuggets from a few ounces to many pounds in weight. Such ores are a tremendous temptation to a certain proportion of those men employed in the mining and treatment of them. A prudent overseeing is necessary to prevent any thieving.

It may appear strange that boys from eleven to fifteen years of age could be entrusted to prevent theft of such material when often it was worth several dollars to the pound. Quite frequently when breaking up the ores, chunks of pure silver were present even in the ore shipments. But we boys had been brought up among the workmen; we spoke Spanish like natives; we understood the mental attitudes and the mental reactions of them. The men had an enormous respect and loyalty for *El Patron Grande* and all the members of his family.

At an early age we boys were entrusted to keep an eye on shipments of ore when they were packed in heavy leather sacks, loaded on mules, and this pack-train made the journey from the mines to the hacienda. There they would be properly weighed once more, this having been done already at the mines. The twenty-five-stamp mill was used for crushing only the high-grade ores. An advanced duty was assigned us not only to act as guard but to assist in breaking up these ores as well as to feed them to the batteries.

Before we had a hospital, when I was fourteen years old, I gave chloroform in a major operation. It was an amputation above the knee, extreme infection had set in and gangrene was working overtime. Alex helped the surgeon, our cousin Dr. Frank Merchant. There was no one else to do it, and it was absolutely imperative if the man's life was to be saved. He was a policeman who had been shot in the leg; the bone was shattered. A native doctor had been treating the man.

He could not or would not be moved from the house where he lived in one of the two rooms. The front one was a small shop where his wife, when he was on duty, dispensed a very insignificant line of groceries such as *panocha,* which is a little block of unrefined native sugar, beans, corn, salt, a bit of coffee, tallow candles, *cerillos,* which are wax matches, a few bottles of *tequila* or *mescal.* This is a native liquor distilled from the juice of the *maguey* or *henequen* plant.

The counter of this place was covered with tin from five-gallon petroleum cans, it was very narrow. We spread a sheet over this, the wife boiled water in a shed back of the store, which was her kitchen. The patient was in a bad way, running a dangerously high temperature. We laid him on the prepared counter. I had a cone made from a newspaper with a little sponge in it, and I gave the chloroform. The surgeon and my brother commenced to work. The antiseptic was carbolic acid. It was not easy for me to gauge the anesthetic. When I got to the point where the almost imperceptible pulse began to get fainter still, I would let up on it. The doctor finished sawing the bone, he was getting ready to draw the edges of the flap together after putting in the drainage tube. From some muscular contraction, the head of the patient rolled and his eyes opened, or at any rate his eyelids fluttered. At that instant the policeman was never nearer to death. The fright it gave me made me empty and awesome amount of chloroform on the sponge and give him a dose of it that would surely have killed him if "his time had come." But *Dios es grande* and he survived.

We had comparatively few mine accidents, but over a period of many years of underground work, breaking hard rock with dynamite, there were bound to be some. In those early years, it was often one of my jobs to hurry to the mines on horseback with the surgeon and rush inside. Or I would help at operations outside as soon as the injured person was brought out. But I never did like it much. In after years, the familiarity with badly cut-up or burned men was increased by the necessity of helping to get them out of difficult places. This was when I had returned from college in the States, though, and went to work underground myself, first as Assistant Superintendent and later as Superintendent in charge of a mine.

You can readily understand that our education in all branches was far from being neglected.

Mr. Clarence Stoner, if I mistake not, was from the vicinity of Hagerstown, Maryland. He had completed his college course just before coming to us and was preparing to study medicine. Like many men who are determined to win out in that great profession, nothing was too terrible for him to undertake if it helped him make the money to complete his medical education.

Maybe Dr. Stoner (I call him that, for he has been a very able M.D. for many years, and I hope is still going strong) might not have undertaken the job if he had known as much about it at the beginning as he did at the end of his tour of duty in the front-line trenches with the Shepherd Infantry. He was possessed of one trait or characteristic which stood him in good stead, which was estimable and much to be desired in the piece of work he had taken on—he was of a serene nature, he took things quietly, never became ruffled, and had infinite patience; hence, at the end of his term of duty with us instead of being a nervous and physical wreck, he had put on weight, and was in the very best of health.

Once a member of the force at Batopilas who was a quick-moving, nervous, incessantly busy person, burning himself up and ruining his health, accused Dr. Stoner of being lazy. The reply of the doctor was characteristic and very true: he drawled, "If I am lazy and if you are not, then I am proud of my laziness!"

He went to California years ago, and "doctored" me once out there in the late Nineties. He was a very successful and prominent citizen of that great State. Taking it all in all Clarence Stoner did accomplish something with us. In the early spring of 1891 Alex was able to enter the University of the South, at Sewanee, Tennessee. After one term in the Grammar School by way of preparation, I managed to pass on to the University, and the next year Conness did the same thing.

When I went to Tennessee for Grammar School, I took with me my *Young People's Encyclopedia, Persons and Places*, and *Common Things*. When I read my first essay before Sigma Pi, the Grammar School Literary Society, it was on the American Indian, and it was copied from the encyclopedia practically word for word. Mr. "Mac" was the Grammar School instructor and the presiding officer at Sigma Pi; he was also house master at Ellmore Hall where I lived. He complimented me on the excellence of the essay. Naturally I felt considerably puffed up with pride. Promptly I went to my room, removed the encyclopedia from my shelf of books, and placed it at the very bottom of my steamer trunk. From that time on the key of the locked trunk remained in my pocket. During my stay in Grammar School I was called on for an essay at Sigma Pi on various occasions; by unlocking my trunk I was able, with the assistance of that book, to prepare "my" essays with the greatest of ease.

Alex and Conness and I loved Sewanee, therefore during our four years there we did not acquire too bad reputations. The long vacations at that place were in the winter, from the fifteenth of December to the fifteenth of March. We were allowed to leave as soon as we had finished our exams and we did some hard riding in order to get home to Batopilas by Christmas Eve.

At the World's Fair in Chicago in 1893, our ores from Batopilas took the premium in their class. The biggest piece weighed three hundred and eighty pounds. We had three of our very strongest pack-mules to bring it out with the regular pack-train and they made the customary forty miles each day. This was done by changing the load every ten miles; and it was a tremendous nuisance. I know, because I helped to bring it out across the trail and I took a week off in football season from Sewanee that fall to see it in place in Chicago. I think I saw it there. But my feet hurt me so badly during that week of sight-seeing at the World's Fair I cannot really remember anything clearly except the state of my pedal extremities.

When I entered the University I was sixteen, Conness was fourteen, and Alex was in his nineteenth year. We behaved fairly well, did not smoke, and drank only sporadically, because we all three were in training most of the time for football and other sports.

Alex left Sewanee in December, 1893, and went back to work in Mexico. Conness and I left in December, 1894. We worked in Mexico for two years, then we returned to the States in October, 1896, with Alex, who was married that fall to Miss Phoebe Elliott, the charming daughter of Stephen Elliott, the Bishop of Texas.

Conness and I remained in the States and took special courses in the school of mines, he at the University of Pennsylvania and I at Columbia University in New York during the 1896-1897 terms. We returned to live and to work at Batopilas in the summer of 1897.

IX

Fourteen Mules to a Stage Coach

In Chaper III I spoke of Jack and Dora's hotel in Chihuahua. That city advanced very rapidly with the coming of the railroad and the increased bullion shipments as the various mines in the Sierras developed. When the ones at Santa Eulalia came into production the shipments of ores and bullion ran into values of many millions of dollars through Chihuahua. This necessitated the erection of all kinds of new buildings, machine shops, a brewery. But Chihuahua was an ancient town and the charm of a Spanish city was never submerged by the usual jerry-building boom.

The hotel of Jack and Dora did extremely well. They were beloved by all those who stayed there off and on. The meals were uncommonly good and there was a greater variety of food than customarily is possible for the price paid. Ham and bacon which were very expensive, due to high import duties, was found on the table always. After the railroad was built, Chihuahua was a divisional terminal and many railroad men stopped there at their hotel.

Before John Worden changed his run to pull the passenger train from Chihuahua south, he was the engine driver on the passenger train from El Paso, Texas, to Chihuahua. One time he got a confidential message as he was leaving from El Paso:

> "The very first thing after you pull into Chihuahua, your tender tanks are going to be inspected for contraband."

The inspection in Chihuahua duly took place; nothing contraband in the way of ham or bacon was discovered. But they did find a false water-tight tank inside his regular water tank, and it was absolutely empty. It is said on the best of authority, however, that the thick smoke breathed by the passengers for a while on the run down to Chihuahua smelled mighty like ham and bacon. This old friend of mine, when I taxed him with his duplicity said: "The story is a libel. I never did in all my life fire my engines with ham and eggs!"

Nowadays when one talks about mines, the mind of the uninitiated turns instinctively to some nationally known colossal companies so vast and so grand that their places on the list of the Stock Exchange or Curb are exactly known, and the eye turns involuntarily to that line on the very long ladder of names. There was a time, however, when there was more romance about mining and more individuality.

Miners in those days were apt to be men who had been in abject poverty, wearing ragged cotton drawers, whose *guaraches*, sandals, were all but gone into fragments of leather, or even rawhide—and rawhide was the last thing in bad form where *guaraches* were concerned; men who had eaten their last meal many a long hour before, whose gourd or canteen was empty of water; men who had no place to return to where they could get another meal unless it was offered by a friend out of kindness. When ready to quit and drag himself to the nearest water, sometimes one of these men would find his El Dorado, his Bonanza, his Open Sesame to a fine house, rich clothes, bullion-covered *sombrero* with large monograms made with his initials in gold or silver, one on either side of the high crown (maybe it would be further ornamented with a four-inch-wide band of "tigre" around the brim), horses, mules, a ranch, good clothes for his woman who from now on was "Doña So-and-so" instead of simply Juana, Dolares, Manuela, or what will you.

I know of no more romantic case than that of Pedro Alvarado from El Parral. The description given above fits Pedro like a glove. He was a *peon*, an honest man, an untiring prospector with that undying hope that keeps them at it year after year. Pedro struck it. With a large assortment of capital letters you can picture only partially how damned BIG he did strike it. Very soon it was not a question of how many hundreds of thousands of dollars Pedro owned, for it went quite a distance beyond six figures.

The ore was "lousy rich" and there was a Hell of a lot of it, and it was easy to get.

Pedro built himself a really big house, an immense pile with a central patio, quantities of costly furniture. . . .

Then as now the salesmen with that unexplainable but infallible instinct which like the graceful bird soaring smoothly in circles alights upon the carcass, swooped upon Pedro as soon as the money, real and undeniable, came in the bank. Pedro liked music, he intuitively liked good music and nice things, for he was that kind of a person. The result was more than one piano; when phonographs were purchasable several were installed in his domicile. It was suggested that Mr. So-and-so was coming down soon from the States on his regular trip, a representative of one of the large silverware manufacturers. Pedro requested that he should come to his home in El Parral to let him look at some of his stuff. The salesman arrived with many imposing heavy trunks, filled with the samples of all kinds of silverware . . . a certain number of each and every design got out by his firm.

A prodigiously long table was arranged in the huge *sala*. Upon this were set out in due order all the samples in the most enticing window-dressing fashion. When this was completed, in were escorted the worthy Don Pedro Alvarado and his good and simple wife. They examined and politely exclaimed over the beauty of the glittering display . . . waiters, pitchers, coffee sets, tea sets, trays, bowls, vases, candlesticks, knives, forks, spoons of every conceivable description in sparkling array. Then Pedro who, when it came to a deal nowadays talked turkey, and hot turkey, spoke up: "*Quanto para todo?*" How much for all of it just as it stands? The fascinated salesman attempted to explain that the display was only the samples, not a complete set of any one thing, but that he could order with pleasure a complete set of whatever design . . . and so forth and so on for a time. But Pedro wanted only what was actually on that table right then and there, and he wanted it to remain exactly where it was to show it to his friends, also and so forth and so on. The bewildered salesman figured it all up and marked it down with a "three-pronged fork" probably, took his money and wired back to the States for another set of samples.

Maybe he sold Pedro the trunks too: they were of no further use to him.

I made use of an expression which may bear explaining . . . "He marked it down with a three-pronged fork."

That expression came into being a long time ago when some wag of a human went to settle his bill at a store in some primitive Mexican village or other in a remote past. . . .

There were many storekeepers who could neither read nor write, nor had they any ideas as to the usual figures employed by the initiated in that mysterious science known as arithmetic. I have known men who did thousands of dollars worth of business, profitable business, whose accounts were kept as follows:

A straight line was drawn across the page from left to right. You came in and purchased on credit one *peso* of goods . . . a circle was drawn half above and half below the line.

Maybe you then bought fifty cents worth of cigars for your friends who had just come in or maybe it was some other things worth fifty cents, *cuatro reales*, or four bits, take your choice. Your account then had a half circle made above the line.

If you bought twenty-five cents worth of anything, you had two lines vertically drawn from above down through the line.

If one *real* worth of salt, for instance, it was one such line.

If a *medio real*—six and a quarter cents—it was a short line just above the horizontal line.

If three and an eighth cents, a short line just below the horizontal line.

Now, if instead of using an ordinary pencil or pen with an orthodox one-point you used a three-pronged fork the result would not be the same, and it would rather favor the keeper of the account and not the purchaser.

Those were grand old days in Chihuahua in the big times. I remember the famous old Turco who many years before had started his career with a trained bear on a chain. Now he ran a monte game with a table a "block long," a bank of a quarter of a million dollars and a reserve of as much again, should it be required. No steel barons ever bucked that game, but I have seen bets made and lost there that no steel baron's press agent need have added zeros to, in order to get a story in our highly conservative press sheets.

Those mining men, who played as a means of relaxation and amusement, went out of the way to keep the facts of their playing from becoming public property; their own crowd knew how much money they made or lost. But they looked upon publicity as bad form. The sources of their wealth were the earth and hard rock. It was not a market for engraved certificates; and what a man did with his own money was his own affair.

As a small boy I used to come into Chihuahua sitting on the driver's seat of a Concord Stagecoach filled with silver bullion, with fourteen mules hitched to it and going at a full gallop. The long whip of the *cochero* would be popping while we swung dizzily round the corners of the narrow streets of the town. We would pull up with a flourish in front of the Banco Minero. The coach I would be on would be the leading stage, and it was followed by several more all of them loaded with silver bars.

The *cochero* with his helper, who was the wielder of the short whip, were of much prominence in their social set. Let us stay on the coach and drive around to the old *meson* with the *conductores.* The old *meson* was a species of inn with stables, corrals, and all the other necessaries. The *conductores* were the men who conducted or drove the bullion trains. Train is certainly not meant to convey the idea of a railroad train.

It needed but a brief time for the admiring crowd to assemble as the stages were driven at a smart trot through the huge *saguan,* that was the gateway at the entrance, while one behind the other they came through to line up in the patio of the *meson.* Back of this *patio* was a inner one where the corrals and stables were located. The throng was lively, good-natured, pleased and happy at the laughing jests passed back and forth. The *muchachas* and the *mujeres* would blush and hang their dark heads when complimented and teased about their beauty and attractions by those "*caballeros* of the pack-mule and stagecoach, who came from the far-away mineral out in the Sierras where were the mines of so-fabulous wealth."

These were "*muy hombres.*" Armed, well-dressed, with money to spend and more than willing to spend it while they were on twenty-four or thirty-six hours' leave in the capital of the State. Let us make the journey back with them as far as Carichic because from that place to Batopilas we ride a mule . . . back across grass-grown plains where immense ranches spread themselves for miles and miles, where thousands of heads of cattle roam; where fine horses and mules are bred by the hundreds. . . .

The mules were smaller than those you see in the United States, but they were quicker, very hardy and sure-footed. With a load of two hundred pounds they would make their forty miles each day for many days on a stretch. To many people, who do not understand, a pack-mule is just that . . . or even less . . . but to us who were totally dependent upon them for all transportation purposes for a long term of years, a pack-mule was more than "just a pack-mule." The same goes for the saddle-mule and for those in harness.

When unlooked-for emergencies arose you had sometimes to ride long distances over very dangerous trails on dark rainy nights. You could loosen your reins and permit your mule to have its head to select and choose its own way for hour after hour. At the end of your journey when you are brought to daylight safe and sound after traveling forty miles from your starting point, your feeling for the mule will be one of respect and affection.

The pack-mule is unlike his brother who makes up the teams for the stages and wagons, and he rarely ever kicks. This is the result of constant familiarity with being handled, because two men always put on his *aparejo* and his *carga.* The men adjust the *reata* so that the diamond hitch will function properly, and this is quite an art. That *reata* is pulled and tugged at from every angle, all slack is taken up to make sure that the load will go safely all day long, up and down hill and along levels, with only an occasional adjustment to take up any little slack that may have developed from the constant motion.

We will take the journey back with the *conductores,* via the *coches* until we reach Carichic. Let us go to the *meson* very early in the morning. . . .

Day is breaking, the driver with his two helpers, who are also the armed guards for each coach, have had their breakfasts—bread cooked by the regular company cook in his cast-iron portable oven. Or it may be corn *tortillas* procured from a near-by *fonda, frijoles* boiled and then fried in a little hog fat, *manteca.* They may have, too, some stew of pounded-up-and-frizzled *carne seca,* which is meat dried in the hot sun.

This stew is highly seasoned and mighty good it is too, particularly when it is washed down with a couple of pints of strong, black coffee.

Before we can take our leave of Chihuahua, we must harness our mules, fourteen of them to each stagecoach.

Each man takes the bridles on his arm for the set of mules it is his duty to bridle, and brings the animals from the corral back into place to be hitched. While we are at this point, let us investigate how these men hitch fourteen fat, sassy mules to a stagecoach and go about it in an orderly well-regulated manner. Two mules are hitched in the wheel as is customary the world over. Every vehicle, as you already know, which is drawn by horses or mules has a tongue or pole projecting outward in the front. The pole is a little longer than the animals.

To hitch the wheel team, one animal is placed on either side of the pole, facing front, of course; his harness is put on him, presuming that he has been bridled first. Next the collar is settled in place, and then you raise the harness, put the hames over the collar, strap them together under the collar; buckle the belly-band. You buckle the breast chain and hitch the traces next to the whiffle-tree; the whiffle-tree is attached to the double-tree; the double-tree is fastened by a bolt to the base of the pole, which is in its turn appropriately and very strongly attached to the front running-gear of the stagecoach.

Now, we have the wheel team hitched, or at least the two animals are harnessed and standing ready to be hitched to the coach. There is a hook at the end of the pole. On the hook is a long piece of good oak timber with a ring-bolt in the exact center, which ring-bolt is now hanging from the hook. You observe two double-trees united to the piece of good oak timber near the end. This means that here we have an arrangement which enables us to hitch four mules abreast, and the four mules are directly in front of the two wheelers.

If we look again at the pole what will we see running

through staples or eye-bolts along the bottom fastened to the front running-gear assembly at one end, which extends beyond the end of the pole, and farther yet beyond the heads of your first set of four mules which you have just put in place? If the daylight is strong enough by this time (or if it is not, then by the light of a lantern or from a fat-pine torch), you will perceive a chain at a proper distance from the heads of your first set of four mules. There is secured to the chain by a stout ring another tree with two double-trees joined to it exactly like the one to which your first set of fours are fastened. Farther along the same chain is a breast yoke to this tree and between it and the breast yoke you back up and harness your second set of four mules.

Once more, what do we see made fast to the chain? Another one of those same trees and at its end a breast yoke. We will back in another four mules and harness them. Now we can hitch the whole bunch, buckle the side lines that are clasped to the two drive reins of each set of fours, and we are ready—after many, many years of practice—to climb on the box, to take up the eight lines and to shake out the long whip.

It will, however, take you a much longer time to become a safe driver of a fourteen-mule Concord Stagecoach than it will to learn to drive an automobile. It will require a greater degree of skill and bodily strength to gather up all eight of those heavy leather lines in one hand to allow the other hand to be free in order to pop your long whip, to pop it with such acumen that you will not miss any spot aimed at even though it should be no bigger than a fly which is annoying one of your animals.

But, you may say, we have hitched our mules before we got them out of the corral!

That is easy if you have trained teams such as those owned by the company, and if you know how to handle them. Otherwise if you walked out into the corral in the dim light of breaking day, your arm full of bridles, and started in to circulate among that bunch of eager, milling mules, you would in all probability take two or more days to bridle them . . . if you were lucky and did not land in the hospital or at the undertaker's.

But if you (and the mules) are trained, you do this: You and your two helpers walk into the corral, each with the appropriate number of bridles on your arms—each bridle belongs on the head of a certain mule to which animal it has been especially fitted so that it will be perfectly comfortable—and you call out in a deep and manly voice: "*Forman!*" This you do three or four times. After a brief interval of milling about, of scurrying and seeking companions and proper mates, the intelligent, well-kept heads of many mules will be facing you. They will all be in perfect alignment, their tails towards the walls of the corral. Each four that belong together are together. It is almost never that you will have to change the position of any one mule in the four. That is easy, isn't it?

We now put the bridles on the proper heads, buckle the throat latches, clip the bits together with the short straps which are found joined at the exact spot, and you are ready to lead your four from the corral. Now that we are harnessed and hitched to the stage, we must drive out of the *patio* of the *meson* through the *saguan* and turn into the narrow street. We must do this without "hubbing" the stone wall of the arched *saguan*.

You crowd your leaders out onto the sidewalk across the narrow street. In obedience to the pull of the right-hand lines, they gradually straighten out towards the right, but they will not take up any of the slack nor do any pulling of the load as yet. The next four mules behind them do the same, and the next, finally you let the wheelers do the trick, thus the old Concord is eased through the *saguan* and they all are straightened out. Then YES, they take up the slack! To the driver's call of "*Mula! Mula!*" they trot off to the accompanying crack of the long whip, which may be cracking only in fun, but which can at any moment raise the hair from any mule that doesn't behave itself.

The company never had less than five hundred mules, pack and stage. We owned many teams of fourteen mules, all perfectly matched as to size, color, and so on—blacks, grays, sorrels, bays. . . .

Now we are off on our trip back to Batopilas. About four blocks, I cannot exactly remember how many—I have walked, driven, marched very many blocks since those carefree, happy days of youth and young manhood—you round a corner to the right approaching the old stone-arched bridge, barely wide enough for the hubs of your wheels to miss the parapet of masonry two hundred years old. On into the outskirts of Chihuahua, through dirt-paved streets lined with one-story adobe houses sometimes plastered sometimes not, you drive into the road that winds for a few miles along the side of a mostly dry *arroyo*. From this, you go into a wider semi-plain where for some miles the round, rust-colored stones and rocks make the stage, strong as it is, swung on several layers of sole-leather straps for springs, toss about like a dory on a choppy sea.

For three hours we travel. Then we see what appears to be a squat square box with a big keg at each corner. This is first perceived dark against the skyline, because you are gradually climbing eight hundred feet since you left the city of Chihuahau, which is about three thousand feet above sea level. As you draw nearer to this box-appearing structure, it is simple to see that it is a building erected for defense. As its name implies, El Fortin, it is a fort, a defense position, where for many years the Mexican Government maintained a force of soldiers to protect against, and to retard the raids of, the Apache and the Comanche Indians. To my great delight, when this tale first began there was yet a considerable force of Mexican soldiers at El Fortin. There were a quantity of mounds dotted about on the landscape, which were pointed out as graves of fallen Indian warriors who had been killed as they would ride round and round the fort, firing at the armed guards with disastrous results to themselves. This they frequently did even in the late Seventies.

As the years passed, we would only stop at El Fortin occasionally for the night. By that time, it was nothing but a prosaic headquarters and mail-stage relay. We would spend the night there when we had reached this place too late in the afternoon to get to Chihuahua before the bank closed, which would force us to lie all night at the *meson* with the bullion. If we were on our way back to Batopilas, we would stop there for the night, if we had been unable to leave Chihuahua until afternoon.

Now across the plain from El Fortin. . . . The mules are now in a lively trot since we are no longer climbing, and they will keep it up by the hour. On the stage route we make no noon stop. On and on until the sun is low enough to strike well into your eyes, then downgrade into the little valley of the Carretas River, where the village of the same name is located on the edge of a free-running stream where the monotony of treeless, grass-covered expanse of plain is relieved by big alamo trees. Carretas was a pleasant stopping-place. We could always procure eggs, and milk and chickens there, chickens were a welcome change from the eternal *carne seca.*

Carretas was near Santa Isabela where the revolutionists years later held up a Chihuahua & Pacific Railroad train and shot down a number of harmless passengers. It was a pretty little town in another shallow valley on the Camino Real, on which "Royal Road" we are journeying with the stagecoaches. Like Carretas, Isabela was also sometimes used as a place to spend the night if the waning day caught us nearer to it than to Carretas.

Now, we have left Isabela miles behind. About two o'clock in the afternoon we are going more slowly, we are traveling among the foothills of the Cusi Mountains. These mountains we circle in order to get around them onto the plain that leads to Ojos Azules, called this because of the Blue Ponds or shallow lakes. There we will make night camp. This will be the last one before we drive into Carichic, where we leave the stage as a means of transportation, having driven one hundren and fifty miles. From Carichic we will take to the hurricane deck of the good and faithful pack and saddle mule.

You cannot fail to have noticed that we are always making very early starts. This we do from Ojos Azules, reaching Carichic at noon. We have the entire afternoon to arrange the loads for the pack animals and to prepare for a daylight start in the morning. Tomorrow we must make a long, long ride. That day's mileage will carry us out from the foothills and well into the big mountains of the first divide that lands us for our first night's camp in one of the most picturesque gorges in the world. A rapidly running stream of sparkling water winds and dips for miles a twisting course between perpendicular walls, bluffs of solid rock. These bluffs are from three hundred to six hundred feet high, and there is but one opening at each side where it may be entered. The openings in the rock are actually very steep narrow breaks in the great bluff itself. The place through which we will clamber tomorrow morning, as we continue our journey, is only a crack, at the very narrowest place about thirty feet wide, and through this a trail has been built. This stream, the Gauhochic, from which the station takes it name, in geological ages past eroded a small area, a triangular niche on one bank, an area of about twenty acres. It is above high-water mark and it has accumulated a certain amount of silt during all these years. It is on this spot that the first day's ride comes to an end, a ride of forty miles.

Our next day's going is over a romantic-looking country to the station of Pilares, small pillars these, Nature's pillars. We pick our way through a series of these pillars for three hours, while we are in the *Arroyo de Las Iglesias.* It is called the

Gorge of the Churches because the bluffs all about us for many miles are filled with columns and pillars and what can well be imagined as church steeples and towers. Most of them are more than a hundred feet high, many more lofty still, and they are blended with entrancing colors where the erosion of the ages has taken advantage of the softer parts of the rock.

We are out of this *arroyo* at last and onto a tableland of majestic pine forests. We travel along this until a clearing is reached in which rests the Station of Pilares. The old unwashed station keeper welcomes us and receives our greetings with pleasure. We clamber down from our faithful mules who are as glad to discard their burdens as we are to be discarded.

The stations are all built to pattern. In good time we will have our supper, and then to our gratefully occupied beds. Those have been taken inside the house by our *mozos*, unrolled from their leather or tarpaulin coverings, and spread out to receive us on the high canvas cots. We need blankets here in this altitude of seven thousand feet even in the warm weather. But in the winter when the ground is covered with snow and the wild wind howls and moans through the pines, you will be glad indeed to pull more than one over your shoulders when the fire has died down in the fireplace.

Next day we have a few more stiff climbs. Then down into creeks and little rivers we go, especially the Eurique River, whose bed at the place it is forded is not more than forty-five hundred feet above sea level. We ride on tablelands and ridges, through magnificent pine forests whose plumy tops reach for the sky until you have an ache in the back of your neck if you try to look at too many of them.

The sun gets low in the west . . . we arrive at La Laja. This station is in the very tall timbers on the banks of a small constant stream, it is never a dry arroyo, of pure cold water. The buildings are on a bare expanse of rock from which this station takes its name, La Laja.

Fourteen Mules to a Stage Coach

Moving on from La Laja the next day after the usual forty miles we come to Teboreachic. We find the same arrangements of houses, corrals, and everything here, except that the area in which they are located is yet more constricted because this station lies in a contracted gorge. From this place we must make an unusually early start in order that we may reach before dark the edge of the Big Barranca, or gorge, down which we will find the Hacienda San Miguel, the mills, and the mines.

X

Christmas at Batopilas

Christmas and New Year's Day were THE big days with us at Batopilas. My father and my sainted mother went to measureless trouble and a great deal of expense to make them so. It was not an easy matter to do this, particularly in the early days. They insisted on a full and orthodox observance of any anniversary, whether it were Christmas, New Year's, or just a birthday, in spite of the fact that our large family celebrations necessitated much greater forethought and preparation for months ahead than would have been required in the United States.

The box canyon of the Batopilas River was devoid of farm products, and everything had to be brought from remote points into the camp, from turkeys down to the ingredients for the mince pies and plum puddings. Turkeys were driven in from the mountains, as many as fifty at a time. They would have to come many weeks before the proper date, because they had to be kept up and fattened to recover from their forty- or fifty-mile hike. Very occasionally some of them would be brought to Batopilas on mules, packed in crates.

My father always insisted on having a suckling pig, and they were hard to get hold of, since little or no stock was raised in that section. The hams, of course, came from the States, as did most of the "goodies" that go into Christmas and New Year's dinners at the family table. It was rare indeed when fewer than twenty persons sat down to table—seven children and my parents, the several gentlemen employed by the company, engineers, chemists, the surgeon—and later on as the in-laws came into the tribe it made quite a gathering.

The other American employees and the foreign and native ones were served at a mess. On Christmas Day the children of the employees, several hundred of them, would come filing in through the big gates down the ramp to the street, which ran in front of the storerooms, which were immense stone buildings. They would step through the iron gates into the garden on that same level, up the ramp to the big patio. There under the orange trees would be laid a long table spread with candy bags, nuts, oranges, cakes . . . these things were to be distributed to each one of those shy little tikes who were too overcome to do anything but whisper *gracias*, and smile.

There was plenty for my mother and sisters to do. Although they had all the native extra help they needed, and fairly good cooks, they were very busy supervising everything. The preparing of cakes and other sweets was a labor of days. The mince-meat, of course, had been attended to long before, as well as the other ingredients that went into these delectable pies. I have already extolled my mother's mince pies. I can hear my father now, going over his wine and liquor stock in anticipation of Christmas preparations and requirements, saying:

"And, oh, yes, one more case of brandy for the mince pies!"

If during the days of Prohibition or even now, the Supreme Court had seen the kind of brandy that went into these pies they would have declared it was in violation of the Constitution that such vintage liquor should be employed in culinary arts.

The wines and liquors, with the exception of the Scotch whiskey, were chiefly secured in Mexico. The Scotch came at regular intervals during the year from Scotland sent by Mr. George Crystal of Glasgow. This gentleman was the largest individual owner and director of the affairs of the Trinidad Shipping & Trading Company, whose steamers plied regularly across the Atlantic through the Gulf of Mexico to Mexican ports, and made their way home through the Caribbean to

Trinidad. The Scotch came in small barrels just the right size so that two of them, one on either side, made a proper mule load.

Conness and I acted once as escorts over the mountains to Mr. Crystal and Mr. L. H. Stevens. We were coming into Batopilas for vacation from Sewanee, at the same time that these two loyal and devoted friends of my father were en route to the camp to pay us a visit. For a Christmas present they gave each of us a beautiful leather case with a silver name plate engraved with our names and the dates. The cases contained a Star safety razor, six blades, a patent sharpener, shaving soap and a brush. Since we were still novices at the dangerous game of shaving, these gifts proved not only our pride and delight, but removed the peril from the ordeal.

There is not one particle of doubt in my mind that those times were the golden age. They were not the horse and buggy days, but the stage and pack-mule age, and there has been nothing fundamentally as fine since those days vanished, whistled down the wind, and the speed mania came into existence.

The apples in the mouth of the baked suckling pig with all of its juicy fellows was brought to us by the Tarahumare Indians. These shy people loaded the fruit in crates of their own construction, and put them on their beautiful small ponies to barter them for supplies or money. The apples had been gathered from the trees which had grown from the seeds brought from Europe three hundred years before by the fearless and tireless men who wore the cloth of the Franciscan and Jesuit orders.

When we were young lads at college and school, our arrival in Chihuahua for Christmas was timed to meet a returning *conducta* if possible, since at that time the stages came into Chihuahua. Later, when the Chihuahua & Pacific Railroad was completed, we went by train as far as San Antonio, which was a way-station on the line, and not to be confused with San Antonio, Texas. The stages and wagons would meet us there. In the first instance it meant a drive of two and a half or three days according to the conditions of the roads. When the railroad was completed this was cut down to about a day to reach Carichic and the company ranch, El Alamo. There we transferred to the faithful mule, pack-saddle and the efficient *arriero.*

The obtaining of Christmas presents was another problem. While the stores in Batopilas carried quite a good stock most of it was of European manufacture. The majority of us were of sufficient age when we came to Mexico to have acquired American ideas as to what a Christmas present should be. The articles purchasable in the way of toys, for example, were out of the question to an American child. Therefore, for Christmas, there always was a quantity of steamer trunks full of Christmas things; the candies generally were of Huyler's make. This was in an era before other candy makers had realized the possibilities of the profits to be enjoyed in fine candies, so that Huyler's was king. The memories of a Huyler's ice-cream soda caused as much water to form in one's mouth then as did the thought of a real "Scotch and soda" highball in the dry days of Prohibition. The trunks were usually packed at the New York offices of the company and shipped to El Paso; or they were brought down by someone who happened to be making the trip from New York to the camp at that season of the year.

Steamer trunks were used because they made a fair load for a mule when one was put on either side of his pack-saddle. During the four or five years we boys were at college in the States, we generally brought them down with us. By diligent travel we could reach home from Sewanee on Christmas Eve. But very often that forced us to make the ride across the mountains in three days instead of the customary five. This was no comfortable undertaking. It was a long hard ride at best when in perfect trim and hardened to the saddle. But to be compelled to double stations and sit in the saddle for thirteen hours on a stretch when we had not ridden for nine months, and to do this during the cold winters which prevailed at five to nine thousand feet elevation, sometimes accompanied with snow storms and high, frigid winds, was no child's play.

Of course you do climb down from your beast every now and then to walk a mile or two and to stir your blood back into circulation; very often you use your Bowie knife to dig the balled snow from the feet of your patient mule, and to "spell the animal." . . . It's not so bad. But at that altitude I do not recommend it if you are on simple pleasure bent.

Once I left Carichic with Marshall Willis, some extra pack and saddle animals, and a bad chest cold. We had to do the trip in two and half days in order to be in the camp on Christmas Eve. We left Carichic late in the day, made Guahochic well after dark, a distance of forty miles, and rested for four hours there. When we hit the *mesa* about four hours' ride out of Pilares (we had come up out of the Eurique valley), I was running a very high temperature and had suffered two shaking, chattering chills. Marshall was greatly alarmed, for my breathing was fast becoming nothing but a sort of rattle. We made camp out in the open. It was bitterly cold. We had the *mozos* scrape back the snow and they built an immense fire. I crawled under a pile of blankets, swallowed another twenty grains of quinine, drank at least a pint of *mescal* mixed with water. I must have slept all of five hours, and awakened refreshed but sopping wet with perspiration. I put on dry underwear in front of the blazing fire, and we started off once more. I recall but little of the hours that followed. We rode, we changed animals when we reached a station, and rode some more and then some more.

At nine-thirty on Christmas Eve we dragged our weary selves into the hacienda in time for all eight of the trunks to be unpacked and the stockings for the children, grandchildren, nephews, and nieces to be filled.

The Christmas days at Batopilas were grand times . . . we even had a tree! This was brought into the camp by men from up in the mountains at Yerbaniz, seven thousand feet high where the pine timber line began.

The Christmas music was supplied by a piano. When I talk about a piano at Batopilas after you have read of the travel it took to reach this place, naturally you are entitled to know how its possession was possible. There were no balloons, and if there had been, they never could have dropped down into that steep-sided canyon which was five thousand feet deep. There were no airplanes and no place to land even if there had been such means of transportation. From what you have read you will realize that although a pack-mule with his *arriero* is capable of carrying almost unthinkable loads, you cannot see him coming along the mountain trails with an upright piano draped across his back.

But don't forget that we still have the Indian.

If you have the time to spare and if you have a clever Mexican who understands the Indian and who speaks his language, and if you leave it to them with no suggestions on your part, that cumbersome box will eventually arrive, and arrive right side up. It is really quite simple.

When it is unloaded from the wagon at Carichic you cut two long straight pine poles. It is better to have them already cut and the bark removed so that they will be dry and properly seasoned. They ought to be about four inches in diameter at the big end. Put one on either side of the piano in its wooden box and tie them in such a way that the weight is balanced with no tendency to be top-heavy when lifted from the ground. I suggest that you allow the Mexican to do the tying of the ropes; they make a very good job of such things. You now have a piano in a box with two poles the ends of which stick out far enough to have plenty of room for at least two Indians to get under the end of each pole; this makes four of them at each end of the piano. They get under the poles in a squatting position; at the word "Vamonos!" they straighten up. The piano is off the ground, and the carriers move off with inward satisfaction of knowing that all they have to do now is to carry this great box for a hundred and eighty-five miles in fifteen or twenty days. There will be at least twenty-four carriers—that makes three sets—and they spell each other every twenty or thirty minutes. They are well fed at your expense if you want a maximum of efficiency. Each man is paid at the rate of a dollar a day. At the end of the journey he takes his "easy money" and trots back home a a hundred and eighty miles or more in about three days, and he has a happy time for some months on his ill-gotten gains!

It is expensive transportation, very. Yes, but think of the pleasure you will derive from that piano. Keep in mind that probably the man who invented the phonograph was a young fellow when this piano made its journey across country and that the voice of the radio was not yet heard in the land. We sang Christmas carols and hymns, gay songs, sad songs, dance music, but you do not require me to itemize the infinite number of musical delights we had with that piano.

On Christmas Day the stamp mills were silent, as they always were on Sundays as well. When I would awaken on Christmas and on Sunday mornings the first thing I would hear would be the cooing of the doves and the brave song of the cardinals in the *guamuchil* and orange trees.

Invariably there was a big *baile* down in the town of Batopilas on Christmas night. In fact there were three dances . . . first class, second class, and third class, *baile de primera, de segunda, de tercera.* Of course we all would attend because there was only the most friendly relationship between the family and the gentlefolk who made up the social life of the town. The *bailes* were subscription dances, committees made all the arrangements, and the expenses were *pro rata.* The ladies dressed in ball gowns and all the gentlemen wore evening clothes. The guests who lived beyond walking distance, like ourselves, rode to the dances on horseback. It was a common sight to see groups of ladies in elegant ball gowns with their long trains looped over their right arms, accompanied by gentlemen in evening clothes, escorted by *mozos* carrying lanterns if there was no moon, clattering along on spirited prancing mounts.

Although we attended the formal dances in long-tailed coats, we dispensed with top hats. We felt that to ride on horseback to a dance in a long-tailed coat was permissible, even though you did discard your holster and belt and wore your gun—which you left in the dressing-room at the ball—in your waist band; but a top hat was a bit too thick. Besides, our horses had never seen us in top hats. When you are all dressed up in a long-tailed coat you do not want a damned fool horse pitching about and disturbing the peace.

I was a very susceptible young man and many is the time I have sent a *mozo* long distances down river where there were a few irrigated gardens, in order to have him trot in with a bouquet of roses for the young lady who at the time was all-in-all to my heart, eyeballs, or whatever organ it may be that is affected under these strange conditions that overtake the youthful *homo sapiens.*

The suppers at the *bailes* were lavish, ample as to spirits and wines to meet varying tastes . . . everything from champagne to brandy. Decorations were plentiful, prodigal in fact, flowers, greens (and these were hard to get), colored-paper streamers. . . . We danced the waltz, the schottische, the polka, the mazurka. Senorita Petra Ramirez with Don Benito Martinez invariably were called on to give us at least one *jota*, and it was well worth watching. When we learned the two-step and the cotillion during our college years, we brought home those dances to Batopilas, which innovations were received more than kindly by the gay Batopilenses. There was no especial hall for the *bailes*. Many of the residents lived in spacious houses which boasted large *salas*. These rooms always opened through one or more doors onto the inner *patio* which was filled with growing plants and flowers of all kinds.

At break of dawn the dance would end. When Alex, Conness and I had finished college and had regular jobs at the different mines, we would have to hurry back home to change our clothes, remount our horses and ride, each one to the mine of which he was superintendent.

XI

Marshall Willis and the Hot Country

Marshall Willis, who was taking care of me on the urgent journey into camp in time for Christmas, was one of the four colored employees who were with us at Batopilas. John Gibson, "Sandy," and Charlie Robinson were the others. John was a sort of handy man, valet, general factotum. When I was a young man on vacation from two years' work as superintendent of San Miguel Mine, my father sent me instructions from Batopilas to procure and to bring down with me a good male cook. I had already hired a young colored man, John Gibson, who had done a hitch in the United States Navy as mess boy. From John I got hold of "Sandy" who had been an officers' mess cook; he was now retired and free to be employed in Mexico. I promptly interviewed Sandy and a deal was arranged.

When the time came for me to return to Batopilas, John, Sandy, and the others of us on this particular trip to camp reached Carretas successfully and spent the night there. As day was breaking we pulled out across the little river to begin the climb out of the valley. Sandy was an excellent authority on all things watery and culinary, but he was unfamiliar with things that roamed the face of mother earth.

He was seated on top of the stage and was carrying a little .32 revolver. When we were about half way up the long hill, we were startled by exclamations from Sandy, more properly they should be called shouts of excitement. Followed loud discharges from his little pop-gun. Between each discharge he vociferated:

"There's a deer! Stop the coach! I see a deer!"

It struck us as decidedly odd that a deer should be so near

the constantly traveled road. However, we halted; investigation revealed to us two little burros in the roadside brush who appeared unduly agitated. One of them had two slight abrasions on each side of that end of him to which his fly-swatter is attached. . . .

We had considerable trouble in convincing Sandy that he had been shooting at a burro and not at a deer; we even had to point out that deer had horns and that burros had only ears. We mounted the stage; Marshall Willis, who was driving, kicked off the brakes, popped his long whip, while between periods of his inimitably infectious chuckle, he urged his team forward up the steep climb of the Carretas hill.

Marshall Willis had joined the force at Batopilas many years before this little incident. He had been seen wandering about Chihuahua, to which town he had come it is supposed, for some very good reason of his own. Possibly the sheriff in some district in Texas along the Rio Grande had a connection with that reason. Before the sheriff had become interested, I believe, there had been an altercation in some dance hall or other between a number of cow-punchers (Marshall had been a cow-puncher in Texas), and fatalities had resulted which made a change of scene imperative.

He knew all about mules and horses, six-guns, and he spoke Spanish as it was used by the average puncher who worked cows along the Rio Grande in Texas. He became an invaluable man to us. For years he had charge of the *conducta.* He handled hundreds of thousands of dollars represented by silver bars; and hundreds of thousands of dollars in currency for payrolls. His devotion and respect for *El Patron Grande,* with very few lapses from the straight and narrow, made him a man who deserved much and who was rewarded accordingly.

This man could ride almost anything that had hair on its back; his shooting with a .44 was the envy of everyone. It was never any bother to secure a chicken for supper when we would reach a station at the end of the day en route with the *conducta.* This usually tiresome piece of work was solved by having the wife of the station keeper indicate to Marshall which ones of her fowls were for sale to us, and he would shoot their heads off with his six-shooter. From this to the frying-pan or pot according to the age of the bird and the condition of our appetites was a question of not more than thirty minutes if we were hungry and an hour if we were not.

To us kids whose idea of being real men was to be able to shoot, to ride, and to rope, Marshall was the greatest of them all. We were given into his care on the trail, around the camp and all. He taught us these as well as many other useful things. We all were perfectly devoted to him and he to us. He took inordinate pride in the prowess of any of us boys, and particularly in the capabilities of Conness. It was Marshall who taught us to sit up on a pack-saddle with a strong hold front and aft to ride the young mules who were being broken to the pack-train. This was done by roping them, holding them down, putting on the *aparejo* and then letting them loose to buck until they tired themselves out or got rid of the *aparejo.* Just as soon as they did that, it was put on them immediately again, and they would start in to buck some more if they wanted to until they were exhausted by the unequal struggle and gave it up as a bad job.

Some years after the death of my father, Marshall took his savings and his bonuses to purchase a small ranch in the Hot Country, where he said he was going to stay until he died. Came the Madero Revolution and the years of unrest and final prostration of our camp. We were all scattered here, there, and everywhere in quest of a living. Marshall was lost sight of entirely.

About eight years ago, the Kid, now a man well past the half-century mark, was working some lead-silver properties not very far from Carichic. When the ore teams got back to the mine from the railroad station in the afternoon on a special day, there was a very old man on one of them. This old man was "Irish." "Irish" was the nickname Marshall had been known by for ages, how or why the Lord only knows.

"Well, I'll be damned," says the Kid, "if it isn't old Willis!" And after shaking hands, "How would you like a drink to wash the dust out of your throat?"

"All right, Mr. Conn. The Governor always asked me that the first thing when I used to run the bullion train and had just got in. It must have been the right thing or the old Governor never would have done it. Lord, Mr. Conn, I am glad to get back to one, at least, of the old-time *Patroncitos!*"

That night many questions were asked and answered on both sides. Old Marshall had been with us since 1884. He had risen by faithful service from cook on the trail to conductor, the boss of the bullion train. He had seen us grow up from little boys to men getting on in years. Many fortunes and many ill-fortunes had been shared in common. This old man was then in his seventies. He heard that one of us was in Mexico working a mine near the old stage road, and he left his home in the Hot Country, saddled and rode his mule five hundred miles across the Sierra Madres to come to my brother's camp. Here he died a short time later.

Marshall Willis had his faults, and his weaknesses. I have never met a man who was worth a damn who did not. But he surely did have loyalty and affection. Unless you have ridden across the Sierra Madres you cannot truly appreciate just what a real good reason there must be to make you do so under the very best of conditions. When old Marshall made his last ride over them, the conditions were anything but favorable, even for a young man, much less for an old one with a very limited supply of money.

This old stager used to say that he did not have to work for a living. "When my father died, I inherited the whole State of Georgia." When the proper interval had passed he would add: "To work in at a dollar a day." And he would chuckle at his own wit. I must have heard old Marshall tell that story one hundred times every year for twenty years, and the last chuckle was just as full of enjoyment as the first.

Charlie Robinson has his own part in this narrative which will come after a while.

It may be well to explain here that Hot Country does not mean Hell. The Hot Country, or Tierra Caliente, was that part of the Pacific coast section of the States of Sinaloa and Sonora which well-deserved that designation. The high mountain and central plain was known as Tierra Fria or Cold Country, and the inhabitants in these sections known as Calientes or Frios as the case might be.

These terms and differences were very marked. A good-natured rivalry—sometimes the nature was not so good—existed; a Frio resented being called a Caliente, and a Caliente resented even more strongly being called a Frio. They differed too in appearance and there were certain disparities in their household customs due undoubtedly to the very clear dissimilarity in the temperature of their respective lands.

I think the chief distinctive qualities between the Tierra Fria and the Tierra Caliente lie in their blankets, saddles, and other horse and mule gear; and the protective garments against the brush worn by the *vaqueros*. The saddle used in Tierra Fria is a squat low-cantled, flat, broad, circular-horned affair, and this horn is like a big toadstood cut flat across, about six inches wide. The saddle has a seat leather-covered half-way forward of the cantle. The cinch and latigo rings are set far forward in front of the rider's leg; the right-hand ring is equipped with a broad double strap, punched for the tongue of a large-ring-buckle to which the cinch is woven. The left side is the same, only the latigo is much longer and is run through and over and upon itself twice or three times before it is buckled.

The saddle is furnished with meager skirts. Unless extreme care is used in the arrangement of the saddle blanket the tree is liable to gall the animal. In order to prevent the slipping forward of the saddle, it is equipped with a cropper whose long ends are passed under the tree between it and the blanket or saddle-cloth. The ends of this are divided into two straps and are taken around the horn and made fast to the latigo ring on either side.

The leg protection against brush consists of *chaparreras* made of heavy tanned leather shaped to the leg and has short narrow straps with many buckles from hip to foot where the *chaparrera* is flared to cover the top and side of the rider's foot.

The headstalls for the animals are not very different in the two sections. In the Cold Country they do not often use the light leather halter under the headstall to which the *cabresto*, or halter rope, is joined. The saddle-cover is generally more elaborately decorated, since the lack of saddle skirts makes it more visible, than is the case with his Hot Country brother. The Cold Country spurs are cumbersome and weighty, they have large blunt rowels, and the spur straps hold it but loosely to the shoe of the rider.

The Hot Country rig is far better-looking and more comfortable, to my way of thinking. The saddle-tree is shaped more like our own Western stock saddle of the better kind. It is frequently adorned by beautifully carved leather skirts, and these skirts are ornamented by tiger-skin *baquerillos* in combination with carved or pressed leather. The cantle is slightly—yes, rather more than slightly—higher than the saddle-tree of the Cold Country. The horn is long; quite slender of neck and small of head, and there is a more graceful curve to the entire tree.

It is lighter in weight than our Western tree. It is made from the wood of the *guassima* tree. This is very tough and strong and light; from this comes the expression "*aggarrar la guassima*," to grab hold of the pommel of a saddle. It is easy on the back of an animal. The latigo rings are set quite far back, almost under the stirrup leather so that the cinch when in place comes just about in the center of the lower side of the animal's barrel. It is a single rig and rarely if ever does a self-respecting rider degenerate to using a cropper. In fact, with the cinch located where it is, and with the knowledge of saddling, it is not necessary, as a Hot Country tree will remain in place under nearly all normal conditions.

The stirrup straps are broad, double, and with sweat leathers for leg protection. The stirrup is made of wood carefully shaped, and with a hood, more to guard the foot against cactus thorns and brush (which are also copiously supplied with thorns), than to keep the foot from going forward through the stirrup.

The Hot Country does not use the *chaparreras.* Instead it uses what are known as armas, which are two tanned, pliable cowhides, correctly trimmed and joined by leather lacings end to end in front, at which point is arranged a hole of the right size. Through this hole the horn of the saddle can project when these hides hang down, one on each side of the animal, in line with the front legs, long enough to reach well below the bottom of the stirrup housing. With these armas, when they have been pulled back and tied at the waist of the rider in the back by thongs, the *vaquero* dashes through any kind of brush and remains unscathed. That is if he has also a stout buckskin or leather jacket.

His spurs are not cumbersome and are very useful. The spur straps are strong and a double chain goes under the foot better to insure its remaining in place; these chains, however, are never tight against the wearer's foot, they hang down from half an inch to as much as an inch at the point of the greatest sag. The rowels are from an inch and an eighth to an inch and a half in diameter with a varying number of points. These are usually blunt; it is very rare indeed to see an animal with a bloody flank or belly. The rowels are set to the spur shank by a strong pin or shaft upon which they revolve freely on the end of the shaft, which is pierced for the purpose. On the outside or right-hand side are suspended two small pear-shaped pieces of steel which jingle pleasantly as the animal moves. They are not, however, only to afford music to the rider and to the animal as you can prove for yourself by trying the following stunt:

Set your animal at a gallop; as you swing down to raise a handkerchief or it might be a silver dollar from the ground, the leg opposite to the side from which you swing down comes up, and the rowel of the spur is hooked against the back of the cantle, which gives you something to hang from. Now, why does the rowel not revolve on its axle and thus break your hold? Because these little jingles have dropped down between the points of the rowel and they jamb against the shank to keep the rowel from turning. So we see that this finishing touch to the work of silver-inlay art which these spurs frequently are, is one of usefulness rather than adornment.

The Hot Country rider has a single narrow strap for a headstall pierced or split, or so arranged that you can slip the ears of the animal through them. This is seldom done. The strap rests on the head of the animal just behind the ears. Under this often very flimsy headstall is a stout halter, loosely fitting with a ring at the back where the halter strap is supposed to be attached. But the Caliente fastens to this his horsehair *cabresto,* which is coiled and which hangs by a leather thong to the pommel of the saddle; when the rider dismounts, this is either tied to some convenient object or the animal is trained to stand when the rope is unslung and the coil laid on the ground.

A *vaquero,* and those others who are much on the trail, will carry two other ropes besides the *cabresto;* one of these is the *reata* (or lasso or *pita,* so-called from *pita, henequen* fiber, from which this rope is made). But it may also be made from plaited rawhide; sometimes, but less often, a twisted rawhide is used. This does not refer to the *reata* used for a pack animal. This rope is coiled and tied to the saddle customarily on the right side just behind the leg of the rider, a little back of the point where the cantle meets the tree. The other rope is the *pial*, which is a short rope worn wrapped about the waist of the *vaquero* and carried only by him. It is used to tie the feet of an animal which has already been roped and thrown.

You may think: "What a lot of space devoted to the description of horse and mule gear. Is that so important?" Yes, indeed! In a country where if you move about at all you must do it either on foot or on the back of an animal, the article upon which you sit, and upon which you live a considerable part of your life and upon the proper adjustment and fit of which depend your travel comfort, is a matter of vast import.

There is another piece of equipment which you will find as useful as the musette bag which we knew to be of infinite convenience in France in 1917-18. This is a morral, which is, in fact, a musette bag, slightly different in shape, made of leather, and it is usually the last thing that goes on the saddle before you mount. This useful article will carry your face towel, your shaving kit, a quart bottle, your lunch, your extra pack of cigarettes, and any number of other requirements for your comfort. Since it hangs in the most handy place possible, from the horn of your saddle, it is an incomparable solace.

If you belong to that section of the human race, you can even fill your quart bottle with water . . . this, however, is considered rather bad form. . . .

XII

Hunting Peccary

When we were at home for the winter vacations from Sewanee my father employed us for all kinds of different work in and about the mines, to give us practical experience. Those months free from academic study were filled with manifold duties. Nevertheless we managed to have lots of fun between periods of actual work.

I think I was riding the horse known as Calamity on a hunting expedition Conness and I went on with some of our friends during one of these special phases in our years of lessening adolescence. I really ought to have been riding a mule, for the trail was very bad indeed. We were going to Eurique; this was the name of the hacienda and home of our Mexican friends, the Becera family. This family might fittingly be called one of the feudal lords of the Sierras. That is really what certain families in the heart of the deep mountains were, but they lived kindly and benevolent lives in those isolated regions, and did not exploit the peons, as was the erroneous idea usually held by the unknowing.

We crossed the Eurique River many pages ago on our first trip into Batopilas. The Becera hacienda at Eurique village was situated many weary miles farther downstream from where we crossed, and it was a hard day's ride up and over the ridge from our own canyon. We planned to go to the Becera Ranch to hunt deer and peccaries for a few days, and were going from there to spend the week-end in the village of Eurique at the fine old hacienda. There were quite a number of us in our party and I furnished the biggest laugh for the first day's peccary hunt.

They had stationed me at a certain post; no one was near

me. I was carrying a Marline Carbine .44, which you probably know has a side ejection. I never knew one to jamb before this particular carbine did that very thing at the precise moment I was trying to have my second shot at a charging peccary. The big old devil broke cover from the tall ototillo grass about a hundred and twenty-five feet from me and he headed directly, determinedly, and very rapidly in my direction. His speed increased when my first shot, fired in a tremendous hurry and a little uphill, succeeded only in irritating his already hot temper by scraping his back. The shell of the carbine did not eject cleanly but stuck in the slot; there I was, so damned scared I forgot all about my six-shooter which was in my belt holster. A pochote tree happened to be very close. I do not remember doing it, but by the time that old boar arrived at the foot of the pochote I was climbing its branches ten feet from the ground. "Now, this is fine," thought I.

My excitement (shall we call it that, in the spirit of charity?) was so intense that I had failed to notice the fact that a pochote tree is entirely profuse in, and totally covered with, spikes accommodated with very sharp points. In my eagerness I had ignored this.

For some fool reason, I still clung to that old Marline carbine and was energetically working on the jambed shell with the point of my knife when I realized all at once for the first time that I was also possessed of a revolver. I knew, even shaking as I was, that I could not miss that old peccary who was valiantly attempting to climb the tree to get at me, and was making circle after circle around it. On his next halt and struggle to clamber up the tree, I leaned over as far as I dared and with the muzzle of my six-shooter not more than four feet from his head, I murdered the wretched beast. It was wilful murder and no mistake.

After he was down—and I knew that he was, for a .44 in the head will stop almost anything—I scrambled from the pochote almost as full of spikes as the tree itself. I appeared to be examining the old boar in an off-hand manner when the Kid and Rafael Becera rode on the scene and remarked with ill-suppressed amusement upon my extreme agility. They had enjoyed a perfect view of the whole performance from the adjoining little ridge where they had been posted in case the wild hogs should select that way down into the *arroyo* instead of choosing my ridge. My prowess as a mighty hunter would have been bandied about those two canyons forever and a day if another and yet more absurd performance had not been staged in the afternoon by Rafael Eureuta. This young man was a friend of our host, and had very recently returned to Mexico after graduating from the University of Pennsylvania. He was a few years older than Conness and I. I must have been around eighteen years old.

We did not linger over our lunch, and the four of us were making our way along the slope of the mountain to a place where the dogs were barking furiously. In a steep-sided little *arroyo* we found, on one side, a cave. Into this cave up an exceedingly precipitous ledge of rock the dogs had chased several big peccaries. They are bad-tempered vicious little beasts. But come to think of it, you really could not blame them for being in a bad humor.

We settled ourselves in the gorge about seventy-five feet from the mouth of the little cave, directly opposite to it. The huntsman climbed the sheer, contracted ledge, placed an armful of dry grass in the mouth of the cave, set fire to it, and waited for about three minutes to see if it would burn well. This was exactly three minutes too long. The peccaries came boiling out of there and the poor devil of a huntsman had to grab the roots of a tree that luckily hung above him. He drew up his feet as high as he could, hanging there by one arm, and struck violently and ineffectually at the hogs with his *machete*. The smoke was stifling him, the flames were singeing the bottom of his pants' legs; to add to his predicament, Eureuta got a bad case of "buck fever" and began pumping his entire magazine of .44's at the peccaries.

The angry animals were thrashing about the dangling feet of the hanging huntsman, and the yells of the poor fellow were enough to wake the seven sleepers, as he implored Eureuta to cease firing. Conness reached for Eureuta and held up the muzzle of his gun when there were yet a shot or two left in the weapon, but Eureuta kept on pumping and pulling the trigger until the missile was empty.

We got three or four of the peccaries. During all this time, they had rolled or scrambled to the bed of the gorge about thirty feet below the place where we were waiting for them. Before we could shoot them for fear of killing the dogs, those wild pigs had disembowled three. They are ferocious fighters, their tusks are wicked weapons. The biggest one of those we bagged that day could boast of tusks five and a half inches long counting the roots. Just the same I cannot help feeling that they were much better men than we were. We were four, each man armed with a six-shooter, a Bowie knife, and a .44 carbine. We had a bunch of dogs of assorted breeds and they were good hunters, much better than we were. Besides this, we had employed several huntsmen armed with *machetes* and an amount of knowledge not possessed by any of the other members of the party, except maybe Becera, our host.

On the way to the ranch that night, we caught two young peccaries by rolling them in a serape as they dashed out from beneath a rock in the bed of the *arroyo*. Two nice little pets we thought they would make. We handled them with greatest gentleness, put them under a big *guacal* or crate. But they plunged about in it all night, going round and round that four-foot square cage. Next morning we went out to inspect them, and they were as dead as Hector. The Mexicans told us it was practically impossible to tame them. They often did what these two little ones had done, "run round and round their place of confinement until they died of *corraje,* died of rage," as the natives put it.

A young peccary is very good eating if the musk sack is cut out immediately after he is killed. This sack is directly over the middle of the haunch. It is powerfully pungent and will taint the entire carcass in a very short time, making it unfit for food. Broiled peccary is quite palatable, not unlike not too fat domestic pig. Notwithstanding their succulence, I, for one, am willing to do without them as a food supply. That was the last time I ever hunted peccaries, or anything else as a matter of fact. Even deer-hunting gave me very little amusement. When my turn came for vacations, later on when I used to go to the States every two years to rest from work in Batopilas, I greatly preferred other forms of diversion and relaxation, like drinking and eating contests in some comfortable New York restaurant, for instance. Although medical men say that deer and peccary hunting are very good for the liver, did it ever occur to you how few medical men have ever hunted peccaries or deer in the Sierra Madre Mountains?

If my liver ever goes back on me I know it will not be from the good food and wines and company I have enjoyed, but from the roasting heat and pumping heart while climbing steep hillsides, nearly perishing from thirst, fatigue, and heat, trembling to such an extent in consequence that I would have had to put the rifle barrel down the throat of the miserable deer to insure a hit. If I had not been compelled to put in so much time fighting spiders, snakes, centipedes, tarantulas, scorpions and the physically insignificant wood tick and red-bug during a hunting expedition, I might have had time to bag an occasional buck.

Pretty nearly the easiest deer-hunting imaginable was at the big mill where the screens were placed across the aqueduct to catch the leaves and other trash, and to keep the nozzles of the turbines clear. The aqueduct was three miles long, and every now and then a deer would come at night to water there. While reaching his muzzle to drink, he would fall into the ditch and be carried down gradually to the screens; the walls forming the sides of the aqueduct were too high for him to escape. At the screens the deer would be found by the man whose duty it was to clear them of trash at regular intervals; with the assistance of one or two other men and two ropes they would pull the creature from the ditch and take him to the place where there was an arch carrying the aqueduct sprung over an *arroyo.* Here they would release the animal to let him make his way up under the arch to the steep mountain side and liberty. No human being could do otherwise than to free the poor beast when captured under these circumstances.

Deer are every bit as destructive as goats. I have more than once rescued parts of the week's wash from deer as well as from those milk-giving quadrupeds. Deer are raised very easily when they are caught young, but they are regular nuisances at all times. A buck can be dangerous too when he has reached maturity. He is able to strike a wicked blow with his forefoot as well as with his horns. We could always get venison when we wanted it. We would give a hunter some .44 cartridges, loan him a carbine and he would bring a deer to the camp for four or five dollars. That was cheap enough in all conscience, since he had to lug the dead deer on his own back for not less than ten miles and frequently farther still. That beats the labor of hunting him for yourself, that is if you are as poor a hunter as I am.

I almost forgot the best hunt that my brother-in-law Ned Quintard and I ever had. It was away up the canyon in the tall pine forests. In that particular spot there is a lake and a very shallow lake it turned out to be, but we did not know that when we saw it the first time. A lot of ducks were swimming about in the water when Ned and I first glimpsed it. We were both carrying Savage Carbines, .303 I think was the caliber. Since the ducks were very tame, being unused to human beings, we could approach them near enough to shoot a brace in their heads. Our *mozos* had gone on ahead of us en route for camp, and must have been a mile or more on their way. In January at those high altitudes it gets mighty cold. In the late afternoon when the sun is almost gone the air grows chillier still, and those damned ducks were two hundred feet from the shore of the lake. To be obliged to ride two hours longer in wet shoes and the thermometer at freezing was not our idea of pleasure. The bed of the lake might be boggy; to ride into the lake where the ducks were floating would be in order, but the water might be very deep. We were about to mount our animals and abandon the fowl, when along through the pines came a little Indian boy about three feet tall herding a few sheep.

He was not so scared of us but that I could entice him near enough to point out the ducks and to show him at the same time two silver dollars. I managed to convey to him the fact that these shining coins were his if he retrieved the ducks. The little fellow grinned with delight. Stooping down he removed from one small brown foot the remnant of a *guarache* and handed it to me to hold for him while he solemnly waded out, picked up the ducks, and waded just as solemnly back to shore again. The water came up to his knees. By this time Ned was rolling about on the ground emitting peals of rocketing laughter. While I had been intent on the negotiations for the salvage of our game, he had been busily engaged in shooting his kodak. He had a perfect record of the entire performance and Ned was almost delirious from mirth for he had incontrovertible proof that would afford many a laugh at my expense when we got home, and there would not be one loophole for a denial of anything on my part.

The photographs were outrageously ludicrous. He had taken them so that the discrepancy in our sizes was more marked than it actually is between six-feet-three inches and three feet.

That poor little Indian boy handed me in all seriousness

that fragment of a rawhide *guarache* . . . his great eyes shone when I gave him the *pesos fuertes*, strong silver dollars. One silver dollar was the price for a fine sheep. His opinion of me for being such a fool *gringo* as to give the price of two fat sheep for two miserable, thin, unpalatable wild ducks the Lord only knows. I do not know what "fool *gringo*" may be in the Tarahumare tongue, but that is what was in his mind I am sure. He dashed off like a streak when he recovered his sandal from me with those two *pesos fuertes* in his grimy little fist, fading out among his sheep in the tall pines.

Ned was responsible for the nickname of "Pud" given me years before this when I was fatter than I was anything else. I mean when that was the only distinctive qualification that I possessed . . . it stuck to me for many years until I grew into the six-feet-three of awkward lankiness which is the only other distinguishing characteristic I have held since.

XIII

Bishops and Football

A visit to Batopilas which caused even more awe and excitement than that of the Governor of the State was the one made by the Bishop of the Diocese, the State of Chihuahua. In its way it was a more notable event, because it meant that children could be baptized by the Bishop and thus go through life with that distinction if with no other. The good people of our camp swept the trail clean and free from stones for a distance of fifteen miles up the valley. There were innumerable arches of green boughs which were not easy to procure although the rains were barely over.

The procession that met the Bishop on the trail consisted of men, women and children, all the population whose duties in connection with the reception in the town itself did not necessitate their remaining at these duties. Forward to meet him and back to escort him fifteen miles each way these devout people tramped; the *Señor Obispo* surely at no other place had received so true and reverential a welcome. The contributions given to him for one good cause or another were rarely equaled from a town of five thousand population. I venture to say that the Bishop had never seen such a well-dressed, cleanly and healthy set of parishioners. That he was tremendously respected there can be no doubt. There is a strong, deeply religious feeling in the Mexican nation which is a characteristic of the people. In our camp many of the men themselves cared but little for attending Mass, some of them did not go at all, but the women were exceedingly devout. They made every effort to instill a belief in the Christ in each of their children.

Their piety was very apparent considering the fact that the

old Padre in our town and district was a man of none too churchly habits. He was endowed with an indolence far in excess of that which any of his flock could hope to emulate. This ineradicable belief, deep-seated as it was, in itself demonstrated the undeniability of Jesus Christ. For God knows I have seen devotion and sacrifice made for the Church under conditions where there was, during many years, an absence of contact with priest, religious training, or church service.

Many of the workmen, fathers of a generous number of offspring, whose wives were consumed with anxiety to have them baptized by *El Señor Obispo*, borrowed the needed money from us for new clothes and many other essentials. We were only too glad to lend it to them so that they need have no feeling of humiliation and the dignity of their families could be properly maintained. Marriages were much more numerous during the visit of the Bishop than over a similar period of time ever before. In fact, it became a perfect mania, everybody got married. Couples who had been devoted and faithful husband and wife, who had reared large and self-respecting families under common-law marriages, embraced this opportunity to climax the successful union as husband and wife by having it recorded that they were married by the Bishop. Many lukewarm swains were swept away by the general enthusiasm; others under the effects of this, and the exhilaration derived from other sources, became victims of certain ladies of vast experience, but with heretofore no yen for the married state. Everybody was happy in the general revival of the favorable occasion to express their faith in the Church of Christ and its earthly symbols . . . the bringing up to date, as it were, of these symbols and customs. Certainly it had a beneficial effect upon the community.

I was home on a vacation from Sewanee at this time. I think I was seventeen years old. Some of the young men attended a special Mass on Sunday morning and afterwards saw the baptism of several hundred children . . . the mothers kneeling on the stone floor of the ancient church; the Bishop and his colorful retinue very dignified; the priest delegated to take up the tickets which each mother carried showing that the preliminary details, fees and the like had been complied with; the consternation and misery on the faces of some poor mothers who had come in from the mountains miles away on foot carrying a little child, or maybe two, to be baptized, ignorant that there were any such details and preliminaries to be complied with. We immediately organized a relief corps. There were hurried contributions of all the cash we had with us, we secured more from a convenient store opposite the church. Jesus Hernandez and Rafael Martinez exhibited promptness and efficiency combined with diplomacy, and the situation was relieved. Those poor mothers were lined up and their children baptized after the first and more provident ones had been attended to. The Bishop was elderly and frail. Even though staggering with fatigue he completed the services which consumed the entire morning and part of the afternoon without a let-up.

This demonstration and many others it has been my privilege to witness in my life, and they have filled me with confidence that an atheist, if it be his desire to convert the world to his condition of non-belief in God and His Son Jesus Christ, must be one of two things, maybe both of them, either a person of abysmal ignorance with an immense amount of time on his hands, or just a plain damned fool. Maybe the two states mentioned are synonymous after all.

The Bishop came to Batopilas and left, and the whole affair was an excellent thing. Those who had gone into debt had to practise self-denial in the luxury of the sums of money they had been devoting to their weekly games of chance and to liquor, or cockfighting and the like.

The ladies in the camp of Batopilas are responsible for the church there not being ruins. For years it had been lacking half a roof, but this was remedied. A large proportion of the funds came from the winnings of the weekly games of poker and other card games at the Hacienda San Miguel. There was no fun in winning money from one's friends unless those winnings could be spent on their amusement. The chief amusement was getting tight and raising Hell; so we decided that all the winnings over a period of a year should be put in a fund for the purpose of assisting the ladies of the camp in the new roof campaign. The new roof was a good one and it did not leak. I believe there was not a thing improper in the way part of the money was obtained. I was one of the poorest poker players in the outfit and I felt very proud that my contributions were of considerable size, therefore.

Bishops were no rarity with us since we had been attending the University of the South.

While I am on this subject I want to describe that seat of learning at that mountainous place. It is located on the Cumberland Plateau, and the name of the small village of Sewanee is so closely allied with the University of the South that the names are interchangeable. Before the Civil War this educational institution was established by leading Southern gentlemen and it was heavily endowed.

We three Shepherd boys went to Sewanee because it had become a place of importance in our lives when my sister May married Edward Alexander Quintard, an engineer and chemist who was a graduate of the University. Ned was the son of Charles Todd Quintard, Bishop of Tennessee. After the Civil War, the funds were absolutely depleted, and no more were to be secured from the original sources. This indefatigable Bishop took up the burden and revived the corpse of what had been a richly endowed Southern university. He raised a quantity of money in England where he was very popular. The University was located near the little town of Sewanee because of the salubrity of that climate and the large grants of land donated for that purpose.

When we went there in 1891 there was a preparatory department connected with the University; it was a military school of high standing. You paid for your demerits by walking extras in full uniform on all holidays . . . far too many of my Saturdays and other holidays were thus spent. In 1891 Sewanee's first football team came into existence through the fact that Alex and Elwood Wilson, who were Lawrenceville men, Elwood's brother Buck Wilson, the artist, Jim Wilder and myself had seen the game played; we had indeed played it before our addition to the roster at Sewanee. No one else there had done so. Nevertheless in the fall we trained and worked to such advantage that we had two games with Vanderbilt and were beaten both times.

But we came back at them in 1892. We took Vandy's scalp twice that year. Out of eight games played altogether we lost one to Virginia. That aggregation of "murderers" outweighed us by about twenty pounds to a man. I was sixteen years old and weighed one hundred and seventy pounds, I played right guard. Penton, my opponent on the Virginia team, weighed two hundred and ten pounds and he was twenty-five years old. We played on a field called Island Park in Richmond. From the center of the field to one goal post was a heavy down grade, the other half was practically level. We held them to 0-0 the first half; we had a very fast light team, and they could not catch us. But when they got us working uphill they absolutely sat on us and held us down. Then they would knock down a few of the line men by well-delivered punches to the jaw, eyes, or nose, push their backs over us for enough touchdowns to run up thirty points in forty-five minutes. When we staggered into the Jefferson Hotel after that memorable game, I said to Jim Wilder our center: "Gosh, Jim, what a face you have got!"

Jim tried to grin and remarked from the corner of swollen lips: "There's a full-length mirror. Have a look in it. That very round, big, yellow and black pumpkin you see in it is you."

That was the year when flying "V's" and revolving wedges were popular, they surely were hard on light teams at best, but when you were against a much heavier team you certainly paid dearly for your sins.

The University of the South when we were there had an academic department, an important theological department, a medical department, and one of law. The University was under the direction of the Southern Bishops of the Episcopal Church. Commencements were in June and were colorful affairs. Sewanee was a summer resort; the students were notoriously good dancers and held many Germans, dances, and hops. The beautiful girls in their summer frocks when assisted by the Bishops, the embryo Bishops, and the laymen who possessed gorgeous hoods denoting various grades of intellectual and spiritual excellence, made the old wooded college grounds and the village streets have the appearance of a flower garden among immense oak trees.

It was my good fortune when we were at Sewanee to know and to hear many Bishops deliver sermons—big men, able men. Bishop Quintard, Bishop Dudley of Kentucky, Nelson of Georgia, Sessums of Louisiana, Bishops Beckwith and Kinsolving. The Vice-Chancellor and Chaplain was Father Tom Gailor, who became Bishop Coadjutor of Tennessee in 1893. There was more than one man in the theological department who became Bishop after graduating. Manning of New York, Craik Morris of Louisiana. . . . Theologues, good pals of mine, were Cary Beckwith who for more than thirty years was Rector of St. Phillips Parish in Charleston, South Carolina, Wilkie Memminger of Atlanta, and many more eminent men in the Church. I do not know whether it may have been their early association with me that later inspired them to rise to such dignified heights or not. However, there is more than an even chance that the desire to be as different as possible from the horrible example set them by my worldly self and others may have been involved in their striving for and reaching the more worthy plane to which they have risen.

While I was at Sewanee, supper time maybe at Bishop Quintard's hospitable board, I have frequently listened with devoted and respectful attention in rapt concentration to one of Bishop Dudley's inimitable stories, when he little realized that I was one of those responsible for the neat whitewashed plank fences in the village being all painted up with "SMOKE T. U. DUDLEY'S CIGARS. GUARANTEED BY THE CHURCH."

I recall sitting on the Bishop's front porch one afternoon during Commencement week joining in on the speculation, when my opinion was asked, as to who could possibly have stolen the skeleton from the medical department and hung it on a wire stretched from the highest pinnacle on Walsh Hall to the top of Breslin Tower. The access to the tower and Walsh Hall roof were so blocked that not until some time after Commencement Procession from Convocation Hall to the Chapel for the Commencement Service had passed beneath the skeleton, could the janitor and his many assistants obtain release of the bony relic, nor remove the two large sheets suspended from its feet on which were boldly painted such informative lines as "GO TO PIGGOTT'S FOR NICE FRESH MEAT". Doctor Piggott was the Dean of the Medical Faculty.

For fear lest the deserving group who accomplished this really difficult feat, perilous in more ways than one, might never be recorded in print, I now after all these years shall give their names: Ned Nelson, Alexander Shepherd, Charlie Van Duser, Arthur Rutledge Young, and John Conness Shepherd assisted by his brother Grant. Thinking it over, I believe that Dump Morris also officiated, for in the words of Darby the Blast "he was in all them things them times."

XIV

Modesty and Other Mexican Peculiarities

I wish it was forty years ago. I would take you into the old Gem Saloon in El Paso, Texas. It was a straightforward bar and gambling establishment with a sort of music hall at the back. There were boxes all around the balcony with curtains. The first time the Kid and I went there we must have been fifteen and seventeen years old respectively. But we were very large for our ages and we ordered our drinks at the bar with a sufficiently sophisticated air to convince the bartender that we were entitled to them.

We really had gone to the Gem to see the music-hall show, but we absolutely had to take one drink and to look over the roulette and faro layouts. That made us a little late in going into the music hall. We walked down the center aisle and an inebriated fool in one of the boxes tried to brain us (or maybe someone else, and his aim was bad), by hurling a couple of beer bottles at us. Unfortunately for him, he missed us and hit the back of the seat in which was quietly reposing a big cowpuncher. This worthy made a few comments best not repeated here as to the ancestry of city dwellers and railroad men in particular, and went directly to the box which the hurler of beer bottles was occupying. From what we heard I judge he came off victorious . . . the offender landed at the foot of the stairs, and shortly afterwards the puncher waved to us as he sat with a quart bottle in one hand and a music-hall girl on each knee.

Those old cowpunchers as a rule were good-humored *hombres*, and carried their poison well. Every now and then when they grew excited they would shoot out a light maybe. But the Lord knows that is a dignified and harmless enough form of amusement if the oil does not ignite and burn the dump down. If you are a fairly good shot you cannot miss a light, even an electric incandescent shows up splendidly over your sight. The *Patron Grande* broke us of this agreeable target practice of shooting at electric lights by charging each and everyone of the lights broken at a very high figure to our accounts. We got so really that an electric light stood not a single chance with us.

The natives thought them almost magical when my father first had electricity installed in Batopilas. The first evening the lights were used, *El Patron Grande* was in the *asogueria*, the amalgam room literally, but it was much more than that, because it was there that the retorting of amalgam, the refining of the silver, and the pouring into bullion took place.

Never in their lives had the natives seen or pictured to themselves an electric light. One of the men was told to go out and touch a button to light up the place. Any other persons might do what this employee did down in the reduction room that first night the electric light plant was put into commission. He got a box of matches and tried to screw off the glass bulb so he could apply a match to the wire. When the glass bulb refused to unscrew he found that it would unscrew from its socket. He did this, and took the bulb over to the work bench, lifted a Stillson wrench—he knew all about Stillsons—and put the bulb into the vise to hold it while he used the Stillson. There was a pop and a crack. He sprang back from this mysterious thing, fell over a wheelbarrow that was directly behind him, and broke his arm! For a long time afterwards he was known as "*luz electrica*," electric light.

The electric lights were a great relief from the former coal-oil methods of illumination throughout the big plant and all the other places on the outside of the mines. They had to be clearly lighted since work went on day and night. The coal-oil lamps, and the lanterns of all sorts and descriptions entailed an infinite amount of time and trouble.

We had large locomotive headlights placed at strategic points. A line of electric lights was run into the Porfirio Diaz Tunnel. They appeared so beautiful to the natives that many women and children used to come there solely to look into the tunnel to see "*las luces tan lindas*," the so-beautiful lights. The throng of admirers grew to be a nuisance and on a certain evening the superintendent said nothing about explosions when the time came to fire a round of dynamite shots inside the tunnel. When these fulminated with a great detonation in the face of the tunnel, the blast of air out at the entrance was so tremendous that the crowd of Mexican ladies had a bad time of it with their voluminous skirts. It was thought to be the height of immodesty to show even so much as your ankle. The women did not wear bloomers in those days. Therefore, the devastation suffered by their modest sensibilities was sufficient to insure their never returning to the mouth of the tunnel again to see the wonderful electric lights. The superintendent did not fail to receive his share of abuse from these outraged females as they took their departure sputtering like setting hens in high dudgeon.

The extreme modesty of the Mexican women was never more apparent than when they were bathing in the river. It was not easy to watch them at this. They carry an ordinary sheet, squat down with the middle of the sheet on top of their heads, spread it out over them, remove all of their clothing, put on an ankle-length cotton slip (all of this is done underneath the tent made from the sheet), then walking down into the river in this squatting position, with the bottom of the cotton slip dragging on the ground about their shuffling feet, they perform their ablutions. No one ever saw so much as the points of one of their toes!

When water is available the people are very cleanly. If their homes are by or near a river of sufficient size they are in the water every chance they can get. The success of the soap factories in Mexico at that time attests to this. The soap came in bars three inches wide, four long, and an inch and a half thick; it was the color of the average home-melted and prepared beeswax and similar in appearance. The laundry women gauged the size of the laundry by demanding, "Tantos jabones?" How many pieces of soap?

Some of the women can swim quite well, but with no stroke that you have ever seen. It is done as differently as possible from the way a man does it, because to be regarded mannish in any way whatsoever was very bad form. They do it the way we kids used to call dog fashion. It is the way most kids in my day learned to keep afloat, but about every second stroke they would hump up their middle sections like a porpoise. Since they are not built on porpoise lines, especially as they advance in years, needless to say they are not as graceful as that aquatic animal.

I used to be bewildered by the way they struck the water holding their arms bent with their hands cupped in such a way as to make a sort of booming sound. They did this a great deal. I heard that the custom originated years before along rivers where there were many bends and bush-grown banks in order to warn men, who might also be contemplating a bath, to keep away. Mixed bathing, of course, was not a thing to be thought of for a moment.

When a woman finished washing her clothes in the stream, she would spread them out in the hot sun over a smooth round boulder to dry, then she invariably took her own bath. Wherever this was possible they put in all the time they could spare from their other duties splashing about in the water.

Commencing long before daylight on *El Dia de San Juan*, St. John the Baptist's Day, June 24, the river in our valley—and in all the others where people were living on the banks of the river—was full of men and women bathing . . . but not together. The booming sound I have spoken of made by the women's beating the water would awaken you even though you slept some distance away.

Among the men there were really some very skillful swimmers. Before we had built the bridge to connect the two banks of the Batopilas River, they would give exciting exhibitions of swimming this stream at high water during the rains, which may have been accompanied by a cloudburst up in the mountains. It would come roaring down the canyon piled up twenty feet and more above low-water mark, running like a mill race, very fierce and muddy. There was but one way to reach the other bank before we had built the bridge, which was to enter the river at a favorable spot a long way above the place where you had planned to land on the opposite side, else you would be obliged to walk there, having been swept a quarter of a mile or more down stream.

At its widest the river was rarely more than five hundred feet across, but to attempt to swim it directly would only serve to exhaust your strength with probable disaster. You had to work your way across very cautiously, gaining distance as you were carried down stream. I have seen a man swim the Batopilas during the rainy season with two dozen eggs on his head tied in a bundle, which was held fast by a napkin under his chin, and not break a single one. On some of the streams in the mountains it was customary to use a piece of chilicote wood which is very light indeed, so light, in fact, that a piece of this wood two and a half feet long and only six inches wide will keep you afloat. A hole was bored near one end on the broadest and flattest side and in this a peg was set, this was thick and long enough to grasp comfortably. You held on to this peg with one hand, the chilicote beneath you, and with the other hand your legs you propelled yourself along. The current would be so swift that boulders of several tons in weight would be rolled under the surface in these unusually high-water periods. When the flood would subside familiar landmarks would have disappeared only to be encountered again hundreds of yards down stream.

A *creciente fuerte*, a strong rise on the Batopilas River, was generally the cause of headaches and extra work, despite the fact that every effort had been expended to make these high waters manageable. The rises came down without warning quickly, and fortunately went past just about as rapidly. But in the short time it lasted, the formidable volume of water coming down the narrow canyon and the precipitous mountain sides could do much damage. It swept everything along with it that was not buried deep in the ground. It meant much extra labor and expeditious work at the hundred-stamp mill—Hacienda San Antonio. What a time it used to give poor old Legs, which was Alex's nickname. He was made manager of that important branch of the business, beginning his work after college. I recall one flood when for fifty or more hours he wrestled with the consequences of one such cloudburst. That meant fifty hours without sleep, notes and messages coming from every quarter, the general trend of which were "When do we get electric current again?" with variations. . . .

Sometimes if the cloudburst was immediately above our unhappy heads, the earth from the mountain side would come down and literally fill up the "ditch," the aqueduct. Then the call would come to all hands, and we would descend upon that mud-filled masonry ditch with several hundred helpers, men with bars and shovels, and we would simply clean it out. It was our custom when we started anything to finish it in the shortest time possible and to do this thoroughly. When you are producing a hundred thousand ounces of silver a month, time means money, particularly if you are paying fifteen per cent interest on overdrafts at the bank. Therefore, we often were at the job for twenty-four or thirty-six hours working in water, and keeping full of energy and joyful spirits by copious drafts of black coffee and mescal. You would be surprised what a devil of a lot more dirt you can shovel when up to your belly in water if you are getting a good pull at the bottle every hour and a half.

The river itself cannot cut deeper, for it is on bed-rock.

Or rather bed-rock is only a few feet below a covering of boulders, gravel and sand. The banks cannot be cut because they are formed by the same bed-rock. The only wide places are used for the streets; the houses and mills were above high-water mark or else they had been protected by retaining walls several feet thick and high enough to be safe above the danger line.

In a country which has a rainy season and where the crops must be planted before the rains begin so that they may get the benefit of every drop of moisture that falls, where the only green vegetables come during that season, every little piece of the river banks where there is a tiny bit of land above the chance of high water, has a brush fence around it and is planted in corn for roasting ears, beans, tomatoes, calabazas, squashes, and usually some melons. How the people did enjoy that change of diet! Some of the older workmen at Batopilas always took a week off to get their stuff planted. When it was ready to be eaten they would move to the *siembra* with their families, and they would camp at their garden for a week or two. Because very little land was available, they sometimes planted their *siembras* as far as ten miles down the river wherever they could get hold of a bit of earth free from other ownership.

During the rainy season the goats' milk, and on the ranches the cows' milk too, is made into fresh cheese every day. *Panela* is a fresh cheese cake six inches or so across and a couple of inches thick. Everybody was happy to be able to get this fresh cheese. With a piece of *panocha*, *panela* forms a delectable dish, a bite of *panela* and a bite of *panocha*. When we were children we thought it far ahead of ice cream, the latter not being available.

Goats were greatly respected animals. Except where we put in irrigation in later years at Batopilas, for nine months of the year there was not a sprig of anything green. But directly after the rains started the goats began to give fresh milk in abundance, and they continued to give it, though in lessening quantities all during the balance of the year. First they would eat the leaves and little twigs from the bushes and brush; then the larger branches as far as they could reach; finally God only knows what they could find to eat but they kept on giving milk. The herds were guarded by a very small child and some little cur dogs. The dogs were put with the goats when they were puppies. After they grew up you could not make them happy anywhere else, rain or shine. They would raise Cain at night if anything approached a flock of their goats, and they would fight to the death to protect their charges.

XV

Two-Legged and Four-Legged Playmates

When Conness and I finished college, Alex had already been at work for at least a year in charge of the Big Mill and Power Plant at Hacienda San Antonio. My first steady job was on the night shift at San Antonio with Alex. Then I was put at San Miguel Mine as Assistant Superintendent for six months, after that I was given charge of the mine. The Kid had his first go at Todos Santos Mine. Then he and our first cousin, Frank Merchant, took over the Porfirio Diaz Tunnel on contract while at the same time they took charge of the Bullion train and ran that. Thus they could acquire change of scene and have a vacation—if you can call riding mule-back three hundred and seventy-five miles, and driving three hundred miles in a Concord stage in sixteen days, a rest. Some persons might call it hard work.

Alex slept at the Hacienda San Antonio while he was in charge of the Big Mill. Conness lived at Todos Santos. He had his cooking and sleeping quarters there, very clean and neat and perfectly comfortable but simple as to furnishings.

At San Miguel Mine I had my home. At the immediate entrance to the tunnel, there was a small two-story building. The lower floor of this, at the level of the tunnel and the track, was used as an office and room where the high-grade ores were stored each night. The second floor I used for my quarters. It was on the level of the road to Santo Domingo Mine, and the building held up that part of the mountain side. I had a plain clean bedroom, on the walls of which hung a gem from the clever pen of Louis M., a comically realistic caricature of old Kemper in the act of braining his daughter with a claw hammer.

This room was separated in two parts by a canvas screen, thus making a dining-room for me where the *mozo,* who walked the mile to the Hacienda San Miguel three times a day carrying my dinner pail, laid out my meals. If they had grown too cold and unpalatable, he would heat them up a bit at the forge in the blacksmith shop directly below. Sometimes I made a change both for the good of my stomach and for my spiritual being. A local woman would provide my meals. She was the wife of one of the miners who ran a little *fonda* where she fed some of the unmarried men.

As a rule we held pretty steadily to some special job that we knew best, yet at times for one reason or another—absence of some one of us, maybe—we had to take over the management of one of the other mines for a period of time. So it was that I had work at various times in each and every mine operated by the company, including the Porfirio Diaz Tunnel. It gave me a fine opportunity to study and better to understand the vein groups and the variety of ores, the difference in these ores; although it may seem strange, in the San Miguel group (all worked through the San Miguel Tunnel) the ores of Veta Grande, Mesquite, Carmen, and San Antonio were very dissimilar in appearance. This was because of the divergence of content and crystallization of the various minerals making up the ores. The ores in the Porfirio Diaz Tunnel group were also distinctive in appearance. To the initiated this variation was very obvious; we could pick up a piece of ore and tell from its appearance which vein it came from, and sometimes these unlikenesses were such that we could even tell from what level on a certain vein that particular piece of rock had come from.

Of course, long before this, the mails were coming in twice a week, and the telegraph came to Batopilas. This made it possible to communicate with the outside world quickly, and we could enjoy this useful voiceless method of contact for sending greetings now and then upon some anniversary occasion to a member of the family who happened to be in the States on vacation or business, or to friends. We could employ the telegraph for hurry-up orders to replace broken parts of machinery and other business matters. We could get all the news bulletins rapidly and promptly when the Spanish-American War came along, much to our delight.

We three sons of my father were burningly eager to go. We were men by that time, although but recently having arrived at that estate. We could speak Spanish like natives, we could ride anything that had four legs, we could shoot accurately, and hardships were nothing to give us any care. But we *were* needed at the mines at Batopilas.

However, our father, who was a personal friend of General Miles, wrote to that gentleman promptly, requesting that he should tell him frankly if there was any real need for us; if he should send us to enlist would we be put in some outfit that was actually going to get into action in Cuba or any other place where our qualifications and training would make us of value to our country? "They are very useful to me," he explained to General Miles, "and it would cause me much inconvenience to do without them. If they are only going to enter some camp to do the manual of arms, I think there are probably plenty of others who would do that quite as well; therefore, they would benefit neither their country nor themselves."

The General replied that there was no chance for a fraction of the men already there to get into action, and that my father would do better to keep us at work in Batopilas. So that is why we were not actively engaged in the Spanish-American War. I believed that the Rough Riders would have been our meat. There was much friendly, though at times strained, fun-making between ourselves and the Spaniards in the town of Batopilas. We were perfectly good friends, and when the news of San Juan Hill, the destruction of Cerveras fleet and Manila Bay came to us, the Spaniards would receive the information from some friends of theirs in Chihuahua at least twelve hours before we did. It would always be the news of a decisive Spanish victory. Immediately a big banquet would be served and they would have a grand time, with speeches and patriotic songs that would absolutely roast the unworthy *Norte Americano*. We would keep quiet till the news of the true state of affairs would come seeping through. Then in our turn, we would celebrate. How I used to regret that we had not been in it!

Gachupin is the term applied by the Mexicans to a Spaniard living in Mexico. He was not particularly popular with the natives. During the days of the War there was more than one disagreement between the Mexicans and the *Gachupines*.

How truly my father understood his countrymen and their politics is witnessed by his remarks that even with the example of incompetence, unpreparedness, and inefficiency afforded us by the Spanish-American War, we would not benefit by the lesson, but when the next war came, would be found in identically the same predicament. That he was correct I personally can testify. I served in France with a regular outfit, the 2nd Division was made up of the Regular Army and the Marines. It was a damned fine outfit. We went through our part of the war with the most inefficient automatic rifle, which was the French Chauchaut. I am alive today, if that is anything to brag about, because I could shoot a hand gun fast and could shoot it accurately. . . .

But that has nothing to do with the old loyal friends we held in high esteem in the towns of Batopilas, Chihuahua, Mazatlan. Manuel Leony, a hundred and thirty-five pounds of courage, wit, humor, loyalty, and ability. The half-brother Angel, that really superb character. Andrez, the daring, long since gone to his fathers. Jesus, Rafael, Anacleto, all the rest of that galaxy of men, whose virtues were far in excess of their faults. My brothers and I live over those experiences, glad and grievous, which we shared in common, work and play. . . .

I can see you now, my friends, as so often in life I have seen you stripped to the waist in a hot, smoking working; the sweat pouring from us as we suffered with fearful headaches from the powder fumes, kidding the gangs along. Getting out the high-grade ores, cleaning out the drifts or winzes or rise so that the next shift could go ahead. Fast workers? People who use that expression in connection with trivial things know little or nothing of what it truly signifies.

On Saturday night I see you in old Don Luis Muzy's, leaning on the counter drinking the usual toast, "*Salud y pesetas!*" or maybe later that same evening, as we made our leisurely way towards the plaza to listen to the band concert at eight o'clock, stopping in Mendoza's to tease Don Benito, that fine young Spanish gentleman. Don Benito, I say, who about midnight would have reached that stage of alcoholic stimulation when he would always insist on trying to ride the black stallion my brother delighted in. This stallion when he was in a mean humor could buck high enough for his rider to see the roofs of the one-story buildings. We would chat on the way up the street with our other friends sitting on the sidewalks in front of their homes in the sultry summer . . . the older women rallying us about some escapade or other, while the younger ones (damned sweet-looking girls) modestly exclaimed, "*Ah, Mama, no digas eso,*" ah, Mama, do not say that! The young ones while strolling round the plaza in one direction as we sauntered in the opposite, would make mischievous eyes at us, or would not, as the case and the nature of that particular daughter of Eve dictated.

Maybe on that same night, or on some other, with permission from the *Jefe Politico,* we would engage the band after the concert was over, to serenade the sundry homes of the *novias* of the men of our party. By this time we were on horseback doing stunts for the amusement of the crowd which would be following us about.

As you know, the town was built on the gradual slope of the mountain which at this point was less steep. The three streets in the center of the town were connected by other cobblestone streets which themselves were so very slanting that thy might as well be called ramps. Precisely in the middle for a distance of a block the main street was quite wide and on this street the church was built. In order to meet the requirements of the terrain the front entrance of the church was six or eight feet above the street level, and steps went up from each side to a platform with a retaining wall facing the street. There was room on this stone-paved platform for two horses to stand comfortably. On the special evening I have in mind my brother from one side and I from the other, while standing in our respective saddles, rode up these steps. When we reached the platform we turned our horses to face the street. The band came close. The player of the bass drum leaned with his shoulders against the front of the wall resting his weary feet and showing the effects of his many potations.

"Jump him off, *Patroncito,* jump him off!" yelled someone in the crowd. This the Kid did. It would have turned out to be only an ordinary show had it not been for the fact that although the player of the bass drum saved himself from being jumped on, he was obliged to desert his beloved drum. The hind feet of Conness' overmettled stallion went through it. Of course, I know that the horse did not actually turn somersaults, because the Kid stayed on him, but for three blocks I never saw anything in my life quite so fast and furious. Fragments of the drum were scattered far and wide. The horse with his rider came up against a high stone wall at the end of the street.

After this exhibition, the *Jefe de Policia,* requested us to go home. The row caused by the admiring throng of half tipsy followers at the hour of one in the morning was more than

Tarahumare Indians at the Chief's house.

Tarahumare Indians running the *Bola*.

Operario and *Peon* with *Zurron*.

A “stop” in the San Miguel Mine.

the *Jefe Politico* could put up with; messengers were coming to him from all quarters of the place with urgent entreaties to "stop the riot."

Poor dyspeptic Don Luis, one of the successors of our old friend Don Jesus Hernandez, was far more considerate to us than we deserved, when I look back on those nights of *serenatas*. We paid but a small sum of money for permission to hold a *serenata*. The fines collected, generally, for disorders were very few, therefore it was of no especial financial benefit to the town treasury. But how the people did enjoy it all! How the *señoritas* peeped out through the slats of the shutters with which the windows were protected, and from inside iron bars . . . with what an innocent expression of their dark, sparkling eyes they would say to you when next you met, "I understand that you gave a *serenata* last Saturday or Sunday" . . . .

Those were magnificent horses, sweet saddle animals that came from Sinaloa. The best of them were from the town of Ahome down near the mouth of the Rio Fuerte. There an Italian, Secannini, of more than usual intelligence and ability, had a big ranch. He was fond of horses and the breeding of them. He imported the finest Morgan stallion money could buy at his price, several thousand dollars. By carefully selecting native mares which showed the most marked signs of the Arab blood that had come to Mexico during the years when the Spaniards brought in good stock, he developed a strain best suited to the climate and conditions with which they had to contend. They were exactly the correct size for the saddle, of splendid conformation, strong, hardy, quick as a cat, and intelligent to the last degree, with gentle dispositions. They were scarcely ever broken until they were four years old, consequently they were long-lived and they never knew the feel of a bit until they were entirely saddle-broke and docile.

A hair-rope halter snug-fitting about the nose was used by

the "buster." After this had been done, the horses were turned over to the *arrendador*, whose duty it was to bridlewise the animal; this man settled him to his gaits and habits of good behavior. It was a finishing school and a fashionable one. When the horse had completed his education there you had the opportunity to purchase something that was, in truth, worth buying; something to ride for pleasure and business; something that had been selected as good saddle material from the herd, and set aside, watched from earliest colthood; full of life and ginger. Something in the way of horse-flesh that could just about talk and had nearly twice as much sense as the average *homo sapiens*.

They ran in color from a flea-bitten gray to a beautiful golden sorrel, the latter usually had a partially blazed face.

In the early morning hours of a Monday as about five of us were riding home from a *fiesta* or a *serenata*, I forget which, we stopped for one last one, that always fatal last one, at a store and bar (all the stores sold liquor in any size package, the size of the package being limited only by the capacity of the customer to carry it). This store was of small dimensions; in fact, when the Kid and I had ridden inside and had made our horses put their front feet on the counter, there was no room left except for a foot customer or two. The ceiling was low and our heads, covered as they were by our sombreros, were bent to escape a bump. We drank our last one, our horses removed their feet from the bar, and we backed outside to commence doing stunts.

I owned a beautiful sorrel horse, that was much more intelligent than I was. I loved to make him stand on his hind legs and walk about prancingly. But this time I was not timing right, when he reached the perpendicular I failed to loosen the rein and he fell over backwards. Being filled with the fatal last one, I was not as alert as I ought to have been and the side of my head came into contact with the cobblestones with which the little street was paved. Next thing I knew I was in my cot at the Porfirio Diaz Tunnel and my head was all wrapped up in bandages. I got busy at once making up a yarn about having hit my head on the roof of the tunnel while riding out on an ore car. As soon as I could manage it, I staggered out to the corral to see if El Oro was injured. He came trotting to the bars at my whistle, whinnying a welcome to prove to his repentant owner that he harbored no hard feelings.

There was nothing "The Golden One" loved more than to dance down the street when the band played a tune that suited him. You did not have to do or to say a thing to guide him. He followed it first to one side of the street and then to the other, keeping it up until the music stopped, all in perfect time with the rhythm. He had the grandest manners for social events! At a *baile* when it was part of a round-up or some such affair where they danced *al fresco* under a big temporary brush roof, you could ride him beneath the roof, drop the lines on the pommel of the saddle, and El Oro would move about the ground where the dancing was going on in time to the music, in and out among the other dancers, without interfering with a soul. He was the great-great-grandson of the Morgan aristocrat I told you about, the one that Secannini brought from the States; his innate gentle breeding and blood showed most plainly when he was in the presence of ladies.

Once I raised together a black kitten, a spaniel puppy, and a game rooster; they were as friendly and as playful with one another as any three animals could be. When meal times came, old Antonio would put their three plates close together and they would contentedly go to work, but woe betide the cat or the dog if either of them attempted to stick his nose in the plate of the rooster! That one would fly at the intruder and administer on the nose of the offender a crack that often brought forth yowls of pain.

El Oro was a great pal of Moods, the cocker spaniel. They would talk with one another in friendly converse. Moods would sit down in front of El Oro and look up at him head on one side, the horse would stretch forth his golden neck until their noses touched. They would stay in that position fooling for minutes at a time. Moods would lick El Oro on his nose; El Oro would let out a quiet snort. Moods would turn away his head as though in remonstrance against the rudeness of having received a sneeze full in the face, and they would do it all over again and again and never move from the original postures. Moods was a funny little dog. If he transgressed in any way and you spoke to him in a low reproving tone of voice, he would creep off under your cot and there he would remain until you coaxed him out with melting tones telling him that you were sorry you had hurt his feelings. Poor little devil!

He used to beg me to take him down to the river to throw sticks for him to retrieve, and he would frisk himself into contortions of delight when you said, "All right, let's get some sticks!"

One of the most interesting characters, in his own particular way, who lived in the canyon was Brad. Brad was an engineer, and a good one, in charge of the only other property being operated in the Camp of Batopilas which was not owned by our company. It was a small development. Brad was a good scout, only he was a bit erratic, doing none too orthodox things every now and then when he had been imbiding freely.

Brad lived at the Santo Domingo Mine, a quarter of a mile above San Miguel Mine on the same side of the river. The road was benched into the side of the mountain and was made possible, as a road, by a retaining wall built up about twenty feet, against which rested the sloping, shingled roofs of some miners' houses.

He was on his way home in the early morning hours. We had all been to some dance or other and were making it back to our jobs in time for the Monday morning day shift, hoping to get an hour or two of sleep before we went to work. When we got as far as San Miguel Mine, Jesus Hernandez, who drank very little, and who was my timekeeper, thought he ought to accompany Brad the rest of the short ride to Santo Domingo Mine, because Brad was a great fellow for going to sleep on his mule or horse when in that state of somnolence, and had fallen off two or three times before. So I said good night to them and went to sleep in short order.

An hour later I was awakened by guffaws from Jesus Hernandez and the night watchman. To my rather peevish inquiry as to why this early morning mirth, Jesus informed me that Brad had gone fast asleep riding along, that his mule had stumbled and that Brad had gone from off his back onto the shingled roof of one of the miners' houses below; through this flimsy affair he fell into the middle of the kitchen, where three women were busy over their breakfast preparations.

This violent awakening, his befuddled condition of mind, combined with his being all tangled up in a pile of wood which had come down with him, caused Brad to think that he was being set upon and attacked. When Jesus Hernandez had scrambled down the wall to the rescue, he found Brad sitting in the middle of the room waving his automatic (the first of its kind to reach camp), vowing vengeance for the insult offered him and challenging the horrified women to mortal combat.

Amidst the screams of the women, the squalls of the babies, the indignant protests of the men who were at a loss to know how their patron had got into their kitchen, Jesus called attention to the remains of the roof, and he did his best to pacify and quiet everyone because a crowd was collecting.

At last they got Brad on his feet and started homewards. Jesus said that all the way home, Brad was maintaining that he would be damned if he would pay for a new roof, that he had nothing to do with putting it there. "By God, if it had not been there I never would have fallen through it. I have a perfect right to fall off my condemned mule if I want to.

By the way, where is this mule? Because if I was down there with those fool cooks, then my mule must have been down there too," and so forth.

All Brad's friends would remind him of this happening every so often by sending him a shingle all wrapped up in a very elaborate package with "Many happy returns" written on it with flourishes.

Notwithstanding all these gay times I'm writing about, don't think for an instant but that life went on at the mines on the regular twelve-hours-a-day working schedule. . . . if you were lucky and did not have to stretch them out to eighteen and even at times to twenty-four. . . .

XVI

The Indispensable Mule

You could buy a saddle-gaited mule now and then in Mexico for one hundred and fifty to three hundred dollars that would make many of the show horses I have seen look mighty dim as far as gaits go. I once owned a saddle-gaited mule that I could never force into a trot even when she was very tired. She would take any horse along at a hard gallop for miles. You could forget all about your reins till you were ready to stop her, and you could do that by your voice alone, if you wanted to. She had a running walk that would carry you for thirty or forty miles as comfortably as a Pullman car . . . much more comfortably than an automobile, for you did not have to think for yourself, nor to be watching out for the other damned fool.

On those horse and mule ranches they watch the young colts with solicitude and choose for saddle animals those that are naturally gaited and also possess other favorable requirements. When the wobble leaves the legs of a young colt, and he is still out on the range, you start him going faster than a walk and if he is naturally gaited he will take to that gait rather than to a trot. If you have competent *vaqueros* you will know very accurately how many of your colts will be available for saddle material.

Our company had two main ranches, El Alamo at Carichic where were kept the stage mules and a few others; and across the ridge between the Batopilas canyon and the valley of the San Miguel River, where there was a good-sized area of foothill with some almost level land, was located El Rodeo where we kept the several hundred pack-mules. This area was formed by a condition resulting from the junction of the three main rivers of the district: the Batopilas, the San Miguel joining it from the east; the Eurique joining these two. They combined, making one river called the San Miguel. Later on in its course, it was joined by the Rio de Oro or Rio de Chinipas, and was known as El Rio Fuerte.

For long-distance riding I personally like an easy trot, but the ideal combination is to have two animals, one of them gaited and the other with an easy trot, and to have them well trained so that they will follow one another keeping together on the trail. Then when you are plum fed up on trot and the old trot-absorption springs are showing signs of re-crystallization you can climb down from your saddle, put it on the pacer and go on for another twenty-five miles all freshened up.

Wherever you have to use mules in the day's work, of course, you must have corrals. Therefore, there were corrals at the two main haciendas. At Hacienda San Miguel were kept some saddle animals personally owned by employees and some the property of the company, also the pack-train of mules used for local purposes. At that place at the end of each month were assembled, shod, and fitted with pack-saddles the mules brought in from El Rodeo to be used for the monthly bullion train. Their stay in the corral at the Hacienda San Miguel was generally from twenty-four to thirty-six hours. In conjunction with the corral at Hacienda San Miguel were the saddle room, the pack-saddle room, and the quarters of the *arrieros* and corral man.

Identification tags were used on all pack-mules; each pack-saddle on the top of the saddle just over the withers had a small brass tag riveted to it, these were numbered serially. A light leather strap went around the mule's neck with a tag corresponding in number. When you had fitted, let us say pack-saddle number twenty to a certain mule, you placed a strap with the corresponding number about the neck of the mule, and that mule used the same pack-saddle for the complete journey. Not until he was to be returned to the ranch at El Rodeo after his coming back from Carichic was his identification tag removed. This saved much galling that is sure to follow an improperly fitting pack-saddle.

If the mule and the driver are to get satisfactory results the pack-saddle must fit the mule who wears it; therefore, on hand we had never less than two hundred and fifty pack-saddles of a variety and size in order that all customers could be comfortably fitted when they appeared for work. There was plenty of work for one leather worker to keep himself constantly occupied; when he was not busy with the pack-saddles he put in his time in mending the *costalera,* which were leather sacks used for all purposes such as conveying high-grade ores, corn, beans, salt and lime. . . .

At the Hacienda San Antonio, half a mile up the river and on the opposite side from the Hacienda San Miguel, were located the Big Mill and the mouth of the Porfirio Diaz Tunnel; out of the latter the ores were drawn by trustworthy mules for many years until the gasoline locomotive was installed. Here was situated the other corral, the home of these good helpers. There was not much space on that steep mountain side, but by retaining walls a yard was established of sufficient size for all purposes, and a considerable portion of this space was occupied by the corral. It was a stone-walled corral, a masonry wall, high and strong, as all the walls had to be of necessity in the canyon if they were to stand the wear and tear. The floor of the corral was cobblestoned with a place left bare for the animals to roll in. This floor was kept clean, being swept twice each day with stiff brooms.

Is a mule smart? I remember one old veteran of the pack-train who had so much personality she was beloved by all the regular *arrieros,* even if she did give them the trouble to pick up her *aparejo* wherever she had dropped it in the corral at the end of the day's march. In the morning you would put on her pack-saddle, and when her turn came you would put on her load. She was ready and willing to leave at or near the head of the train, be it a small one or one consisting of a hundred or more mules. She was a black mule with a Roman nose, and was fifteen years old when we retired her. You could always locate her, for she never failed to be at or near the head of the pack-train all day long. But she was absolutely sure to be the first mule into the station enclosure at the end of the day's travel, right up front too, at the edge of the portal to have her load removed first. Until the load was taken off she was as patient and well-behaved as any mule could be, but once the load was gone, instead of waiting with the others to have the saddle removed in turn, she immediately sought and found a corner or a corral post and she would get the edge of the *aparejo* against it. Then she would heave and draw in her belly and slip out of the saddle, although the cinch had been plenty tight enough to carry her along all day with no danger of any loosening in the load. She would give herself a vigorous shake, find a suitable spot to have a gratifying roll, shake herself once more, close her old eyes and go to sleep until feeding time.

This story of mules and another and yet another I was telling to a Know-It-All American who came down to Batopilas to have a vacation from the Wall Street atmosphere at the same time I was returning from a vacation in the States to go back to work.

We watched one old saddle-mule attempt to hide away each and every morning. When the man delegated to saddle him, would enter the corral, the mule would lower his head and slowly slink into the farthest corner of the corral behind the biggest group of mules to hide. But when he would be discovered, he would consent to be bridled and gave the man no need to throw his rope in order to catch him.

After the noon halt to tighten and adjust the packs and pack-saddles, I showed the fellow the mules that would immediately start out to make their way to the lead at every spot where the trail was wide enough to pass those ahead of them.

I pointed out to him five or six special mules who had been the last ones to be loaded and the last ones out of the corral in the morning, which were the first ones always to be in the station enclosure at the evening halt, standing where they would be the first to have their packs removed. There was fierce competition between them to keep the lead, they would race each other wherever there was a level stretch on the trail in order to accomplish this.

I explained to him that once a new mule had made the trip there was never any danger of his trying to leave the trail during the long forty miles to the next station, because this new mule knew that on reaching this place, the load would be taken off, he would be watered and fed, with at least a rest of twelve hours. And I told the K.I.A. American and I showed him much, much more that had come to my knowledge and observation through childhood's and manhood's close association with mules, upon which sagacious beasts the life of the mining camp of Batopilas depended.

And then the poor fool told me that a mule could not think. That only those creatures that walk about on their two hind legs can think. So I picked out a lively young pack-mule for him instead of his usual saddle mule the next morning when we set out. I had him saddled. The man did not know the difference until he got in the saddle and the mule walked about on his hind legs till the rider fell off. After he had recovered from the surprising conduct of what he thought was his gentle saddle animal, and I had dusted him off, his regular saddle mule was brought to him and he climbed aboard. He found that the old reliable moved forward as usual at a nice, easy, ladylike trot. But he still maintained that mules do not think. He could not realize that the pack-mule would rather have its two-hundred-pound pack than to carry a less heavy rider; be free to go along as he pleased eating a bite of grass now and then; drinking from each stream he passed through; having an independent time, than to suffer the martyrdom of being pulled and hauled about by the bit, even though he did have a lighter load, but a load that slipped and slid about on his back for seven hours. It never entered the head of the creature that walked about on his own two legs that the mule was more than willing and anxious to put forth the extra exertion to throw his rider, in order not to be compelled to carry something on his back that he was not familiar with, not to have that damned bit in his mouth. . . .

Like Hell a mule does not think! That pack-mule did exactly what any thinking human being would have done rather than to be obliged to bear that dumb two-legged animal about on his back.

There are many more human beings or two-legged beings than there are mules; unfortunately this discrepancy in numbers is increasing constantly; hence, there are few cases where a mule has been known to run to the doctor. Besides, the doctor does not keep the same careful records.

At El Rodeo we had a ranch-keeper and there after each *conducta* were sent back the pack-mules who had just completed their round trip to Carichic with the bullion, and the return with what we may call the "fast freight." These mules rested here for at least a full month, since there were a sufficient number of them to have an entirely fresh set of mules for the next *conducta*. Now, mules *will* get galled. While every effort was made to see that they were kept free from any risk of disease, still blow-flies and other sources sometimes produced an infection on some part of their anatomy that spread rapidly, and which would soon, if unattended, cause death.

El Rodeo was not less than twenty-five miles by the trail from the Hacienda San Miguel and it was a steep trail on both sides of the ridge. More than once one or more of the old-timers finding themselves suffering from a bad sore that was spreading would decide to come to the corral at the Hacienda San Miguel to undergo hospitalization. They would arrive at the main gate of the hacienda and await admittance about midnight or early morning, having naturally preferred the cool of the evening for travel as against the more fatiguing heat of the day. They would be admitted by the night watchman at the main gate, a big wide gate of iron bars locked by a chain and a massive padlock. All he had to do was to open the gate, and the mule or mules would quietly walk in and make their way to the corral accompanied by the night watchman who would open the corral gate for them, then notify the corral man that a mule or mules had arrived from El Rodeo in need of medical attention.

But it was not always need for hospitalization that brought them to the corral at Hacienda San Miguel. Some of the mules belonged to the group that were used to do local work during the year, taking supplies to the various mines and bringing down the first-class ores. These mules were fed corn each evening, along with all the hay they could eat. Their terms of duty might be from four to six months when they would be sent to El Rodeo for rest and grazing. Several of the older mules liked their corn so well that a few days or weeks of grazing satisfied them, and the longing for succulent grains of corn would bring them back to the corral at Hacienda San Miguel. There would be little or no use in sending them back to the ranch again, for they would be back at the corral very soon after the man who had taken them returned. This would mean two days' time for the man who had acted as escort.

There is more difference between two particular mules than there is between two particular human beings. Unless you appreciate this you would do better to go home and stick to railroad trains. They may be of opposite color, size; come from ranches miles apart, but be as alike in disposition and in habits as any two things can possibly be. If one mule makes friends with another mule, the two will be positively inseparable. If this is the case, it is preferable to take both of them with you on a trip, or you will find that if either one is able to escape the first night, he or she will beat it to the place where the other one is. Of course, you will be able to take your journey, if you leave them apart, but you will know that you are riding a very unhappy mule.

I told the Know-It-All American about the mules, about fifty of them, who were used in the Porfirio Diaz Tunnel's eight thousand feet of single track. These mules made two round trips and then went to the corral for a rest. Each driver had his own set of mules. They hauled into the big tunnel two ton-and-a-half empty side-dump cars, which were loaded at the chutes and came out of the tunnel at a fast willing trot. There was enough down-grade for drainage purposes, which made the cars come out with little or no exertion on the part of the mules.

Now, when a mule had made his two trips he went to the corral, as I have said, for his regular interval of rest. When his turn came to go back to work in the tunnel he was willing enough to go in to do his "trick" and he would do it agreeably. But just try to make him do more than two tricks . . . or rather please do not try to make him do an extra trick. Because the work must go on, and there must be hauled out hundreds of tons of ore and waste every twenty-four hours. If you should try to use force in order to compel the mule to take a third trip with the assistance of ten or twenty able-bodied men and the expenditure of two hours in time, you might be able to accomplish your object. But the loss of time involved, and the possibility of the hospitalization of a man or two *and* the mule will hardly compensate you for the effort put forth by all concerned and the experience gained.

Got enough about mules? I'm sure you have, but don't think that I am not going to talk a great deal more about them. Damn it! I love 'em!

We know very little about the usefulness and power of man-power as such, when unaided by mechanical power. Suppose a thousand feet or so of wire cable one and a quarter inches in diameter on one of the big spools they wind it on at the factory for shipment is needed in our mining camp. It is ordered; the railroad deposits is at the appropriate point on their line; you get it loaded on a wagon, and you haul it to the end of the wagon trail. Now what? Either one of two things takes place. If you use man-power, a number of men unwind the cable stretching it along the trail which they expect to traverse. At reasonable distances from each other the men station themselves, put their *serapes* on their respective shoulders. They then get under the cable and walk with it to the destination, reaching there not today nor yet tomorrow, nor even the day after tomorrow, but eventually. As a matter of fact, if those men whose duty it is to have laid out the work are alert as they ought to be, the "eventually" will be just exactly the correct time for it to arrive.

If you use mule-power in place of man-power, you start at one end, making two small coils of the cable; at intervals of about fifteen feet, you make two more coils, and this operation is repeated until you have all the cable unsevered in any place made up into a series or a succession of two coils or rolls, each set of two coils connected one with the other a distance of three feet apart. The coils are made of sufficient cable according to its size so that no single coil will exceed a hundred and fifty pounds in weight, thus making each group of two rolls weigh approximately three hundred pounds.

The *arrieros* line up their pack-mules, one mule opposite to each two coils. These are then placed over the pack-saddle, one roll on each side and are made fast by the *reata* and the diamond hitch. All the animals are thus loaded, and there you have a gigantic snake of mules as the string thus connected mince their way over the trail.

The first gasoline locomotive ever used underground in that country was employed in the Porfirio Diaz Tunnel, a tunnel drilled through the hardest damned rock that ever was drilled, a grand piece of work. The gasoline locomotive was a queer-looking piece of machinery. A Mayo Indian ran it, he listened to the name of Amado Bobolla, or "Molly." Molly was the nearest the old German mechanic Bill Schwender, who came down to Batopilas from the factory in the United States to set it up, could get to pronouncing Amado Bobolla.

I imagine to the everyday person in these times who needs a piece of machinery, and who has the money to pay for it, the buying of it and the getting of it to its final destination is a simple matter, even though it should weigh tons. But if you have confronting you the problem of setting it up in a canyon in the Sierras it has a different complexion. A mule load is three hundred pounds and the bed-plate of the engine, for instance, weighs several tons. You ask the manufacturer to have this bed-plate especially cast for you in sections not to exceed three hundred pounds each, and to have them so arranged with lugs that they can be bolted together. That is simple, is it not?

"Yes, but my God, the expense!"

Surely that is true; and you want to be very certain that you have a good proposition that will earn money enough to take care of that additional expense. If you are contemplating any such business in the mountains, be sure to select a man to do your calculating who has had to climb over and through many of them, preferably on the hurricane deck of a mule.

If you want a mule to stand perfectly still, blindfold him. For this purpose and reason each *arriero* carries a *tapaojo*, eye cover. This is a strip of leather about five inches wide and twenty-four inches long, with a loop in the center through which a finger is thrust for convenience in carrying. At each end is attached a double leather *correa*, which makes the lash. When this is knotted at the right point, the strap placed over the mule's eyes with the thongs drawn back behind the ears of a mule, and on its neck, it successfully blindfolds the animal. This blindfold is used when shoeing and so on. The lash is also an unanswerable argument when the animal shows symptoms of lost inclination to moving forward. . . .

We had an American employee who was an elderly man, very circumspect. Came time for his vacation. As often happens, his closest friend was a young fellow, a nice enough youngster, but just that . . . a youngster. Maybe he did have something over the normal amount of the spice of the devil in his make-up. Our old friend had an exceptionally good saddle-mule which he left in the care of the young companion when he went off on his vacation. The mule was ridden with exceeding regularity after dark by the youngster; in fact, so consistently did he ride the animal up a certain narrow street of uncertain reputation to stop at a certain house, that the mule formed the habit, of course, of going to that one particular place at fall of night or any other time.

In about two months the owner returned, and saddled up his mule one evening to ride down town with the rest of the crowd. The mule was true to habit, but he showed a singular lack of judgment; he turned at the fateful corner and despite the rider's most persistent efforts took him to the certain little house with the adobe wall around it, stopped and waited for him to dismount. The beast was indeed so very obstinate about remaining before the house that by the time the animal was convinced that it was not the desire of the rider to dismount, nor even to tarry in that vicinity, a throng of people had collected. Meanwhile the lady of the house posed in an enticing attitude in the gate of the adobe wall, and made uncomplimentary and sprightly remarks about the *viejo asqueroso*, the disgusting old man.

Before the evening was over, we inveigled the youngster into buying the faithful mule he had trained so well, for considerably more than *el viejo asqueroso* had paid for him two years before when he was a perfectly decent creature who kept to reputable streets. What irked the oldster most was the fact that he had been paying attention to a good-looking perfectly respectable young widow. There is no one who relishes a risque tale more than a Batopilense, and he was horribly afraid that the story would be handed on to her.

He was in such a state of relieved anxiety and consequent elation that we had made the trade and the youngster was now the possessor of the mule, that he had his saddle changed immediately to another animal and rode past his true love's house on Tom's little prancing pony, Chupa Rosa—which means humming-bird—solely to establish an alibi. Then he bought wine for the crowd, and took such a quantity of it himself that the widow came near getting him tied up with a promise of marriage before he could get away from her.

XVII

Vacation Journeyings

We have wandered far from El Camino Real, far from the Royal Road, where a paved highway now takes the tourist, artist, writer, to the capital of Mexico at a rate of speed we thought only birds could attain when you and I started out on this journey. Before I die I must go back to see, via a railroad train, that west coast . . . see El Fuerte, Culiacan, Altata, Mazatlan. . . . I wonder if the shades of my former friends and their faithful mules and horses with whom I have traveled those hot distances will see me and some night speak to me. I would like to take you to Choix on that trip to the west coast, where more than thirty years ago I danced all of one night, drank beer, *mescal, novalato* rum which they called cognac . . . danced until breakfast time, after I had ridden eight solid hours the day before. After breakfast I saddled up and rode sixty-five miles more to El Fuerte and climbed aboard a mail stage on my way to Culiacan.

A pleasant cool spot was the hotel in El Fuerte kept by old Dona Concha. Her hostelry was a hollow square with a portal all the way around four sides. The courtyard was a big *patio* filled with lime and orange trees, planted with many flowers of all colors and perfumes.

I enjoyed a good bath in my brick-paved room. There were some big clay *ollas* with wide mouths from which I had my *mozo* dip huge gourds full of the cool water which he poured over me for a shower. When I had dressed I sat at a small table in the corner of the portal to sip limeade with something else in it; at this place Don Francisco Torres joined me. Don Francisco Torres, now forty-five years old and getting fat with a prosperous forwarding-agency business.

For many years Don Francisco had been bookkeeper at Batopilas as a young man. When he went to *El Patron Grande* to tell him that he much regretted it but he would be compelled to leave us because his sweetheart was a young lady of his own home town El Fuerte in Sinaloa, and could not separate herself from her parents, and in order to marry her he must move there to make his residence, he was congratulated and given a handsome wedding present. My father promised him the company's business in the forwarding of freight for the purchase of corn, beans, lard, onions, great quantities of salt used in the reduction of silver ores by the then popular chlorinization process; leather in immense amounts used in making the *zurrones,* and other bags for the packing of the high-grade ores; rawhides (for the sulphides of silver after being sewed up tight in canvases were finally sewed into bags of well-wetted rawhides, which when dry it took an axe to open, and in which covering they went all the way to Aurora, Illinois, there to be smelted down and brought to sufficient fineness to be marketed by the New York Office in the New York market) and many more articles that a company of such size employing over fifteen hundred men requires in the way of things that life and business demand.

That is how it happened that Don "Pancho" Torres and I were having our refreshments together with an affectionate reminiscence of times past, discussions of the future. . . . Lord, what days!

There was I, thirty-eight hours without sleep, having ridden a hundred and more miles and danced all of one night during that time, accepting an invitation to a *baile* for this night and planning to take the mail stage to Culiacan in the morning. Were we physically fit? God! If I could only live it over again under the same conditions. Young and strong and well-to-do; knowing the people, loving them; able to do any reasonable thing; when under the influence of liquor and a fine horse, a smile from a lovely Sinaloense, able to do many more things that were far from reasonable, by the non-initiated pronounced more than foolhardy. . . . Ah, well, an old cripple can at least dream about those halcyon days.

The *mozo* at Dona Conchita's is sleeping when I roll in from the *baile* at Don Pancho Torres' home in El Fuerte. I have eaten much, drunken much, danced much, told many pretty girls that I loved them . . . which was more than true, I did love all of them. But the mail stage will be pulling out at four A. M. I change into my riding clothes, because nothing else will withstand the dust I am going to encounter during the ride to Culiacan. With a cup of very strong, black coffee inside me, with my *mozo* and saddle-bags and small steamer trunk, I set forth for the *meson* from which the mail stage is to start. A rather weary-looking moon dips down to the horizon, is drowned bit by bit in the Gulf of California a short ride towards the right. I am feeling mighty good . . . in love with the world. . . .

"*Buenos dias, señores. . . . Señor Padre, tambien es madrugador,*" good morning, gentlemen; Señor Priest, you are also an early riser.

I am happy, I take pleasure in making the disgruntled other passengers exert themselves to respond to my cheery greetings. There are only four of them so far, and I thank God for that. For during the coming hot and sweating hours which are to be my dusty fate in that rocking stage, four are plenty. The Padre, of all the passengers except myself, is the most affable. He exclaims:

"Whom have we here? My good friend Don Pancho with others."

And as I live, sure enough here comes Don Pancho Torres with some friends and a couple of *mozos* carrying bottles. With merriment and good feeling we all have a drink . . . two or three drinks including the *cochero* and his helpers. These worthies are called aside by Don Pancho, he tells them what a Hell of a fellow I am and what a friend of his and the like. This means later on if other passengers are picked up on the road. I will be given the seat on the box and will not have the body of a stranger, hot and smelly and sweaty, sitting on my lap.

"*Vamonos, señores!*" let us go, gentlemen, says our coachman. We scramble in after one last *abrazo* and hand-shake; the brake is kicked off; with the "*Mula! Mula! Vamonos!*" and the cracking of the whip we are away from El Fuerte and heading south.

What the other travelers did for the next few hours until we arrived at the Remuda Station in a small town where breakfast was taken, I do not know, but this individual slept like a top. When we went to the table, you and I, we did full justice to the *huevos al ranchero* with plenty of good hot *chili* and coffee to stimulate the digestive apparatus. When you have dashed some cold water over your face and head, dried them the best you can on a tiny clean towel supplied by the *fondera*, you are ready to clamber aboard the stage with fresh mules all hitched and waiting, to bounce along towards Culiacan. There *con el favor de Dios*, with the favor of God, we will arrive tomorrow afternoon.

That part of our trip was uneventful; very dusty; none too comfortable; but who cared? We were catching up on lost sleep; we were on our way with plenty of money; after two long years of very hard work underground in the mines what we craved was change.

Culiacan was not a bad town. We would have a day and two nights there to play about in and to become acquainted with new people before taking the narrow-gauge railroad to Altata. At Altata the coastwise steamer would pick us up to take us to Mazatlan where we were to embark for San Francisco. A pretty, nice little city was Mazatlan. We made Culiacan on schedule. After seeing about getting our baggage to the hotel and giving the *cochero* something to insure him certain relaxations, into the shady *patio* through the wide door we go. You put your name down in the book, you order up an extra amount of water with which to get rid of the dust too thick to mention; you climb the deep imposing stone stairway to the second floor; and you are shown into a clean, spacious, almost-cool room.

Ah, but you are glad to wash and wash and wash again! Glad to change into clean clothes. You can hear a band playing somewhere not to far away: the sun is going down. When you have had a glass of beer you are ready and anxious to go out to look the situation over. As you reach the *patio* there is an attractive woman cutting tube-roses. To your "*Buenas tardes, señora,*" she responds and introduces herself as the wife of the proprietor. This lady has a sister who lives in Batopilas, the wife of Don Torfilo Orduno. Of course, you know him quite well and you now recall that you have a letter for her. You hasten back upstairs to your saddle-bags to get it out; with that accomplished the *señora* hastens away to read the letter to her husband . . . these people are very clannish, very fond of their relations.

The rattle of the dice says that the *cantina* is over on that corner of the *patio*. That is the way to head for . . . cold beer has not been enjoyed for a very long time. The dice box is being worked by two Americans, apparently. One of them is the biggest man you ever saw, the other is quite small with a jovial red face, the kind of face that would not tan although its owner should spend a hundred years in the tropics; it simply persists in peeling, peeling.

"Jim Coley," the big man, "at your service."

"Patrick Flynn," the small one, "at your service."

The former is from San Antonio, Texas, and by birth a Texan with a capital "T." The small one is from San Francisco. One is a salesman of pumps, big pumps like himself. The other, a son of Erin, as good an Irishman as one ever knew, is the local ore buyer for the Selby people. Both of them are full of fun with a proper capacity for the wine when it is red and the ability to carry it. A pleasure, they said, for them to have an addition to the American population, hence, a welcome like that to a long-lost brother.

"I thought you said you were the only Americans here. What's that sick-looking fellow playing poker over there using a Mexican deck of cards, with the two *gachupin* card sharps?"

"Him? Oh, yes, I forgot about him. He's a poor unfortunate lunger, came down here for his health. Looks to me as if he was slated. This makes the third day those two have been taking him along. Hell! It ain't right. Reckon I'll be doing what I intended to do last night when this damned Irishman took me out to supper and got me loaded. Drift over, feller, after I get in the game, and watch the fun."

This we did. We witnessed one of the cleverest exhibitions of card manipulations I have ever seen. Coley bought chips; the Spaniards were glad of what they believed to be another sucker. The game progressed, and Coley began losing and overbetting his hand, apparently not liking it from the way he talked. The betting grew heavier in about an hour. The older of the *gachupines* suggested that the limit should be removed. When buying his chips the sick American had shown a number of bills of large-appearing denominations; in his wallet there must have been nearly five hundred dollars. I thought Coley agreed rather reluctantly to the suggestion of no limit; the sick man shrugged his shoulders, and muttered; "Hell, it might as well be this week as next!" and he agreed. The poor devil was surely a sick-looking man to be going about.

Pretty soon, Coley and the sick man began to win. Every time the deal came to Coley, he, the sick man or one of the others drew tremendous hands, too big, I thought. I caught the eye of the Irishman, he elevated his eyebrow and winked at me; then I knew that Coley was in command of the situation.

Just one hour was needed after that for him to be in possession of all there was on the table. He threw down his cards. Once or twice one of the pair of card-sharpers had started to say something, but his partner had shaken his head. His own skill with the cards was such that he felt himself to be in command till it was too late.

"Now," said Coley to the American, "how much money have these blacklegs taken from you, you poor damned sick fool?"

"I have lost about five hundred dollars."

"You lost it all right, but it was stolen," commented Coley counting from the money that was before him five hundred dollars, and handing it to the poor fellow. One of the Spaniards sprang forward with abusive language. Firmly I pushed him back in his chair:

"Wait a moment. Easy now on the rough talk, this gentleman has some more to say. If you are *vivo*, as all Spaniards like you, of course, are, you will take your medicine and keep quiet. You are being addressed by the grandson of the man who invented poker, and he can twist your neck like a *pollo*, too, if you get him roused."

"Now, you mining man," said Coley to me. "You can talk this language better than the pilot of Christopher Columbus' flagship. Just you tell these men here that many years ago I put away as being useless the old tricks they tried to pull on me. Tell them that every hand I dealt has been dealt crooked. Tell them I'm going to give back to this sick man what was his. I'm taking what I put in the pot in chips, and they can go, and I don't care how damned far they go so long as they don't come fooling around here again while I'm here. What is left is theirs."

It was done as the big man commanded. Those two humiliated Spaniards got up with their money and left. I followed them to the door and when one of them turned to utter a threat vehemently, I opened my coat and pointed to my *faja* where my Colt .45 was visible, being thrust there as was my custom in the evenings.

"Little man, I do not know who you are, nor where you come from, as for that I do not care. But look at this and remember I am from the mines in the Sierras."

The supper was more than usually palatable because of the letter with news I had brought to mine host from the family away off there in the deep mountains. Coley told us an interesting yarn as we ate the meal and when you hitched it up with the occurrences of the afternoon, it shows that you cannot know too much about any useful thing. Even when Coley was only sixteen years old, he was a big man. It was easy to pass himself off as over twenty-one. He got himself a job as a brakeman on the Santa Fe. He worked for years with the same crew and conductor on freight trains. In those days much more often than now trains had to lie on sidings for hours at a time, especially freight trains; there was no such thing as regular short runs.

Time had to be passed somehow, and the conductor, who was possibly the most skillful and dishonest card player in the Southwest, according to Coley, taught him every trick and crookedness that was known to him. By constant practice Coley became entirely proficient. His immense but supple and quick fingers made the manipulation of the cards simple for him. When he was twenty-eight years old, he left off playing cards for money. Although he tried his best to fight off the temptation to cheat when playing he discovered that the habit was too strong for him, so he quit playing for money. He kept in practice by playing now and then with his special friends and making monkeys out of them "in the effort to save as many innocents as possible from the curse of being sacrificed to the professionals they would undoubtedly come into contact with in life's journey."

Coley had a grand sense of humor, and he was a fine upstanding man, honest and kind. He said he had rarely found his match or superior and he took pleasure in getting into a game with a crooked professional. He loved to make him squirm and sweat and hunt excuses to leave the game when he realized that against himself he had an equal.

We joined forces to make the trip to Altata and Mazatlan.

The narrow-gauge railroad that ambled from Culiacan to the port of Atltata was the proud possessor of a little train which carried five passenger cars, made by converting a flat car into a shed with rough board sides and a roof; there were glassless openings at intervals for windows. The little train was taxed to capacity pulling all the flats it could move on the morning we took our departure for Altata.

A *fiesta* of several days' duration had attracted many attendants at Novalato. Novalato was a few miles from Culiacan and the crowd had come there to see the sights as well as to visit friends or relatives, before returning home. Their time was up and they took the train to drop off at their home villages along the road to the coast; some of them were going as far as Mazatlan. These residents of Mazatlan were readily distinguishable by their more cosmopolitan clothes, the rouge of the women and the unmistakable appearance and manners of the men. The country women wore the universal cotton frock appropriate to the climate, easily washed; skirts very full, stiff with starch and smooth with ironing. The starch was made from a bit of flour and water; the very small iron had been heated in all probability over a meager few coals, for wood is very hard to come by in that coastal, cactus-covered region.

At noon we reached Novalato, where there was a big sugar plantation accompanied as usual by its mill, where the excess was converted into what the West Indians call "ron." The Mexican or the Sinaloanse do not call it rum, they dress it all up in a dignified bottle labeled with various and sundry high sounding titles, the most popular of which was cognac. It was at this place, at Novalato, we came upon the source of the liquor with which I was familiarly acquainted in the Sierras as cognac.

Coley took me with him when he went to the office and warehouse of the company where the proprietor filled a basket with assorted bottles to fortify us for our coming sea voyage of about fourteen hours. He handed Coley a draft on a house in Mazatlan for a large sum of money. Coley, as I said, sold pumps and he had just installed for the Engenio (sugar company) a set of pumps with capacity so great that they pumped all the water from the river into the irrigation ditches. This was the reason for there being a great deal of sugar and hard liquor in that section. In fact, the result of the drumming of Coley's pumps was frequently felt in the heads of those persons living hundreds of miles away, and being tasted by those whose coffee had to be *con azucar*, with sugar.

For lunch we were guests of Don Tomas, who held the train an hour in order that we might be able to finish the meal in comfort. It was to this foresight of his that we owed our seats in that conveyance, because he sent four peons to sit in the passenger coach to hold the space until we were ready. Otherwise we would have had to hang on to some projecting plank or something else because there were at least twice as many passengers on the little train as it could possibly hold. We made the regular stops, and a few additional ones to pick up those who had fallen off, or been pushed off; finally late in a sweltering afternoon we reached the coast and were lightered out to the smallest damned little steamer I ever saw permitted on the waters of the ocean without a nurse.

Our captain was a handsome young Mexican in immaculate whites and quite drunk already. By dint of much cursing and yelling in which he was echoed by all the crew and most of the roustabouts on the lighters, interspersed with references to the ancestry of all concerned extending back many generations, we were squeezed on board, the anchor was raised, and we steamed away across the bar. The last half of the sun was full in our faces.

Considering the size of the ship there was an uncommonly spacious saloon. When we passengers all were forced to cram ourselves into this on account of a sudden violent storm, which came upon us about ten o'clock that night, we could appreciate what a slender bit of room a human being can live in when he is threatened with the possibility of being washed out to sea. It was the very devil of a storm; soon all were saturated, the waves broke over the bridge, destroyed the glass in the windows of the saloon, or the ports or whatever it is you call them. There were screams from the women, curses from the men, howls from the children, cries of hope, of encouragement, of despair. . . .

We had a clear exhibition of the time needed to slip from curses to prayers. When we would be struck by a particularly heavy sea, drenching everybody afresh, prayers and promises for penance filled the air. In the confusion resulting from fright many persons had thrown themselves on the floor or been dashed there willy-nilly. Amidst the struggles of the others to get down onto their knees to pray, I found myself on all fours with two weighty arms around my neck nearly choking me. This state was greatly exaggerated by a quantity of the very cheapest and loudest perfume ever concocted assailing me. The roar of the wind was as summer zephyrs compared to the hoarse bellowing of that old Madam, who had some younger women with her, and who was pouring into my ear a cataract of prayers to all the saints in the calendar. Coley was still standing like a rock, his back against the wall, his pillar-like legs braced with what seemed at least six women hanging to him. This became visible when as suddenly as we had run into the squall we ran out of it, and the brilliant moonlight flooded the interior of the saloon. Of course, we rolled and tossed about for a while, but it was possible to disentangle ourselves and to get out on deck.

"This," observed Coley, "is a very small but a damned fine ship; any ship that with this mob on board can turn over four times and come up with no more water in her than she now has is a *boat*, young feller, and don't make any mistake about that."

"Want to make me acquainted with your girl chorus? How many do you usually hug at one time?"

"Before this, I have lost pocketbooks. I certainly was not going to give any of those little things of pleasure the chance to get my wallet and escape with it in the confusion and darkness. A couple of them gave indications recognizable by a man of experience of not being entirely overcome by fright. The acquisitive instinct, my boy, is very highly developed in some Christians, and from what I heard inside there I believe that each and every one of them was a very good Christian. Did you ever hear such a thorough knowledge of the calendar of saints? Get out your corkscrew, feller, here is one I saved. . . ."

We landed at Mazatlan only two hours late, and found that beautiful little city all arrayed in gala attire, for it was the "Saint's Day" of the Governor tomorrow and two-days holiday had been declared with bull-fights, balls for all classes, *primera, segunda, tercera.* This is a very good way to give *bailes;* there are no fool heart-burnings, no hurt feelings; no pretensions as to the class in which you belong and the corresponding ball you attend.

I presented myself to our agents and bankers, Hernandez, Mendia Y Compania, cashed my drafts, and met the head of the house. He told me that he would send to my hotel invitations to the *baile* for that night and tickets to the special bull-fights on the morrow. This was to be a very distinguished affair, in that there were to be a professional quadrilla of bull-fighters and for three of the bulls the *picadores* would be *caballeros* of the district who would ride their finest horses; and besides this there was to be a roping contest and exhibition at the end of the *corrida* which should prove very interesting.

We thanked this hospitable gentleman, hurried to the hotel, unpacked our trunks to get out our evening clothes to be pressed while we spent the rest of the hot afternoon watching chicken fights. They were about like all other chicken fights, except that these were interstate and the stakes in consequence were bigger than customary . . . anywhere from one thousand to ten thousand dollars. It is pretty nearly as quick a way to win or lose that much money as any I know of; that is, the way Mexicans fight chickens.

I'll have something more to say about chicken fights later. . . .

XVIII

Vacation Journeyings Continued

In Mazatlan could be procured the very finest vintages of wines and liquors. When we have enjoyed what might be your choice in these refreshments, we shall go to the Plaza. The municipal or military band is a good one, it plays before supper for an hour or so, and afterwards for a long while. Beguiling girls and women are sure to be walking around in one direction, with their delicate swaying gait, while the men stroll around on a separate path in the opposite. Thus you may admire without offense if you do not make the observation too apparent. You will see some blondes, many with hair of a coppery sheen, and others with that blue-black hair that is so fascinating. In Mazatlan are families with Spanish, German, English, French strains, combined with some of the Indian blood which gives perfection of form, feature, coloring, and grace of movement not to be excelled anywhere in the world.

Being inclined to be lazy, although having had but little opportunity to indulge this taste, and also tired, because Coley and I had been obliged to stand up nearly all night on that wretched boat, we found a comfortable bench on which to revel at our ease in this exposition of melody and beauty. We smoked perfect cigars made from tobacco grown in the Valle Nacional near the town of Vera Cruz on the Gulf of Mexico. A delightful form of entertainment is that custom of music on the Plaza, which is prevalent all over the country. Even the smallest town or hamlet has a band or orchestra; at least twice a week, there is a *serenata* at night or late in the afternoon. I shall not dwell too long on these remembrances . . . the longing to return to those places and those times grows too strong for me the more they are lingered with in memory. . . .

A well-proportioned ballroom, splendid music, well-gowned and beautiful women, who dance the waltz, the schottische, mazurka, with a perfection of achievement. A Spanish gentleman and his graceful partner, when the floor is cleared, dance a *jota*, a rhythmic delight to see. You are introduced, you go to supper, you drink champagne, you dance some more until nearly daylight. But you will not see a man, and God knows, never a woman! show any signs of inebriation. Getting drunk in the presence of ladies in those days in Mexico was not considered good form.

How we did sleep after a cold shower and a rub-down! We nearly overslept, in fact, which would have caused us to miss the bull-fight. This was the best show of its kind that I had ever seen. The *picadores* were mounted on very good horses, not one beast was so much as touched by the bull, which was the original intention in a bull-fight. The *quadrilla* was capital. Maybe it was Cara Chica who did the *espada* work, maybe Ponceano Diaz, maybe any one of several of those dexterous men. I do not recall accurately, so many years have drifted by . . . but it is a fact that he gave a supreme display of his art. The two bulls chosen for the exhibition of the gentlemen *picadores* could not have been better. I confess to shivers of apprehension when the two *caballeros* rode in on their perfect saddle animals, one an iron gray that made your heart swell with admiration, the bay no less beautiful, and admirably well ridden, both of them.

That was a sight to hold the eye and to thrill the mind. A charging bull. The horse head-on like a flash to meet him. The rider with the set lance catches the great bull in the exact spot. There is a straining between them, the pain in the shoulder of the bull forces him to try to back away from it. The horse follows, follows, keeping that even pressure on the lance, side-stepping as the bull does so in order to prevent that animal from slipping from off the lance. The exciting, screaming crowds, shouting "*Viva! Viva!*" At last the game bull, brave, bewildered, exhausted, with stiletto sharp horns, gives it up as a bad job.

Of all bull-fights that kind is the best for skill. Bull-fights vary much, as do baseball and football games. I remember one in Chihuahua that for a demonstration of the brutality that a crowd on pleasure bent may develop exceeds anything I ever saw.

The entertainment in Mazatlan was filled with color and enthusiasm. The Governor of the State was present, his family and his staff. Many hundreds of the representative people of the city and surrounding countryside were there . . . bewitching women old and young, gayly dressed, handsomely dressed. There was a great deal of money, of really substantial wealth, in this part of the Mexican Republic at that time. Some of those present were descendants of *Hidalgos* who had migrated to Mexico from Espanola, and direct from Spain just after the conquest.

How those young *caballeros*' sweethearts felt when the crowd in *El Sol*, the uncovered part of the grandstand, and in the more distinguished *Sombra*, the shaded part, went wild at the exhibition of consummate art and skill of the perfect horses and fine young riders, I do not know, but I remember clearly what a tingling sensation it gave me.

The last act of the show was very good too, although a sort of anti-climax. Three young bulls were let into the arena, full of the lust of life. Two boys twelve and fourteen years of age rode in and roped the pawing, bellowing beasts in every conceivable way, by any one foot, by the horns, whatever combination they chose . . . throwing, tying.

As a token of my appreciation I sent two cases of champagne to the donor of the tickets, to be delivered to him after I had sailed for San Francisco. It was as little as I could do under the circumstances.

On our way out from the bull-fight, at the foot of the stairs, we heard a woman's cry. Looking back I saw an elderly lady pitch and fall forward. Both Coley and I jumped for her; we caught her before she struck the steps below. She was a heavy lady, and we saw that her ankle was injured; we offered to carry her to her carriage.

"Let me have her," said Coley; he picked her up followed by the grateful husband and the family, large and small. We made our way to her carriage a block away. Coley deposited her safely therein. Amidst the exclamations of "*Que fuerza!*" and "*Mil gracias, señores!*" we backed away, sought for, and found a hack in which to drive back to town.

"What did the lady weigh, Coley?"

"What do you think?"

"Damned near two hundred."

"Just about," agreed Coley; and he had carried her with little or no effort.

"What do you weigh, Coley?"

"Oh, two-twenty-five."

"How tall are you, Coley?"

"Six feet three and a half."

"And all muscle," added I. I was no weakling myself, but this was just a fast whale of a man.

Before bathing and changing our clothes, we were sipping an iced drink when we received a call from a nice little fellow, son of the rescued lady. He brought a message, an invitation from his father with excuses for his non-appearance in person, again to thank us for our kindness and would we "honor his poor house by joining us at dinner the next evening, a dinner given to His Excellency the Governor?" This invitation we *did* have to decline; the good ship *Peru* sailed at noon and we could not delay our journey. Thus ended our stay at Mazatlan for that time.

I think it was on this trip to San Francisco that we came into the Golden Gate only a few hours after the *Rio* had gone down on Mile Rock with nearly a total loss of passengers and crew. It seems very strange now when we have the daily news on shipboard via radio, to recall that the first news we got of anything that had taken place since casting off was when the pilot clambered aboard to take us in at the end of the voyage.

San Francisco was to me the most cosmopolitan city in the United States, but I did not linger there this time, remaining only just long enough to have some clothes made and then quickly went to Evansville, Indiana, where I officiated as best man at the Kid's mariage to Miss Sarah Clifford. It was a gala affair, and they left among the acclaims of the party for a honeymoon in New York and Europe. I also continued on to New York, not with the bride and groom but a day later. Some of the fellows in the wedding party were old Sewanee men, Henry Soper of Henderson, Kentucky, in particular. After we put the bride and groom on the train there was a lot to talk about . . . talking is a dry business if you are mixed up with several southerners, one of them at least a Kentuckian. I will not swear to it, but I think I caught the train for New York the next night.

The habit of wearing long-tailed coats stood me in good stead during my eagerly awaited vacation in that city of delight . . . New York. But it did take me a while to get accustomed to the top hat which I was not used to wearing in Mexico. A top hat is something that I like very much, but it is a "hair shirt" when one has been away from it for two or more years. The first evening of my arrival I hastily adorned my brow with one, and promptly forgot all about it, dashed from my hotel in a hurry to keep a captivating appointment, and sprang into a hansom cab. Naturally my forgetfulness of the topper resulted in its hitting that part of the cab entrance especially designed for the discomfiture of "farmers come to town." It was then too late to do anything about it except to work if off my ears and thank heaven that those necessary appendages prevented its going any farther down on my neck. In the dim light afforded by the cab the article of apparel did not appear too utterly impossible after I had straightened out the telescopic effect and smoothed its ruffled fur. Only when I entered the hall of the place where I was expected to dine, and the dignified butler wearing that frozen martyr expression with a dash of virtuously restrained condescension peculiar to butlers, took the hat from me, did I realize why the crushable opera hat had been invented, and I cursed the individual who inaugurated in that particular era the fashion that an opera hat was not *de rigueur* for all occasions after dark, and that the silk topper was the one and only thing for the theatre.

My long legs made it necessary for me to buy an aisle seat whenever that was possible. This helped the silk topper a lot. I always forgot to remove it from its rack where it nestled beneath my seat when someone wanted to pass in or out . . . and always at the most inopportune time. My silk top hat was kicked and dragged about by the trains of stout ladies for a distance of half a row of theatre seats to wind up in such a condition of undisciplined furriness that it proved a source of augmented income to Knox. But that was of little comfort to the owner who had to go about in it for the remainder of the night. When I say the remainder of the night I mean that literally.

If you have been away two years or more in a mining camp in the Sierras of Mexico, and you have thirty days' vacation in New York City, the hours there are entirely too precious to waste in sleep, you can do that on the Pullman going back to El Paso. What a grand thing it all is to remember! An enlivening musical comedy with some especial one in the cast; supper afterwards at Rector's or Shanley's or Martin's or any one of a dozen of the old places where there were such honestly good things to eat and to drink; the final satisfaction of breakfast at Jack's which was the third hearty meal in about nine hours, but which always was devoured with gusto. . . .

These memories have helped me to pass many a weary interval while sitting on a rock by the side of the track in some wet level two thousand feet underground, with a great hunger and fatigue upon me, not having seen daylight for twenty-four hours. There were many emergencies that presented themselves in our lives in the mines that kept us out of bed for three days or more. I often think when I have heard persons complain about the discomfort of staying in bed, how easily they could make it comfortable by simply staying out of it until they were so very tired they had actually to drag their feet. When I hear someone complain: "Why, I was too tired to go to sleep!" I wonder. That condition is not one of being really tired, it is just the first stages of weariness. . . .

But I certainly was not tired when on the way back to Mexico I ran against Coley in El Paso at the Hotel Sheldon, where I was stopping for a few days before continuing to Chihuahua. It was fun to see each other again, and very late that night we dropped into a small barroom on a side street to have a last one before turning in. When we entered, we noticed that the bar was deserted except for a lone individual standing at the far end, facing the door. He caught my attention because of his set attitude, which relaxed when he had scrutinized us. The barkeep moved to the other end and regarded us inquiringly. He beckoned us to move to his position instead of waiting for us to choose our own places at the bar, and spoke to Coley by name. He produced the poison ordered and conversed about everything in general in a jerky fashion. Finally he said softly to Coley:

"You better get out of here, Mr. Coley. There's goin' to be trouble. Hell will pop around here if the sheriff comes and he is sure to do it. They want that bird over there," with a jerk of his chin towards the lone individual. "That is Blank." He wiped his bar nervously. "He knows damned well he'll swing as he deserves, and he'll fight it out. He is goin' to mess my place all up sure as Hell."

"Bad egg, eh?" said Coley. "Well, let's have another." He raised his voice. "Come on over here and join us, stranger."

"To Hell with you and your drinks! I can buy my own drinks. And what's more you ain't goin' out o' here ahead o' me to tip off any—sheriff!"

This *bribon* was on my left. I was standing facing the bar and sort of leaning on it with my elbows. But when he started to get nasty I slipped my hand down and took hold of my gun, which I always wore in the front of my waistband when I went out at night. I had this bad egg covered under the edge of the bar, the overhang as it were, and he did not know it till I told him about it.

"You are a bad man, eh? Keep your hands where they are, I've got you covered. Put 'em on the bar. Quick, I say! You are not dealing with a tenderfoot. Now, Mr. Bartender, give this bastard a full glass of liquor and so help me if he don't drink it down, I'll turn him over to the cops. They can take him to the coroner. Frisk him, Coley."

Coley did this with great rapidity and thoroughness, producing a short-barreled .44 which he carried with him to where we were now drinking beer. I was mighty relieved I had not been obliged to shoot the fellow. The sheriff came in; he had been looking through the window and had seen Coley disarm him. The sheriff handcuffed the offender and took him away, thanking us. This bad man was a product of San Francisco and had done-in a defenseless man on the waterfront in that place. He had made his get-away from there to El Paso, where he had expected to cross the Rio Grande into Mexico. Only this very morning, he had murdered a helpless, unarmed section boss up the road towards Sierra Blanca, and had been forced to come back to El Paso.

Coley and I left the little barroom and had some ham and eggs at the Chinaman's before turning in. But I demonstrated again to Coley before we bade each other good night the advantage of a woolen *faja* as against a belt when you want to tote a gun in the only place *to* tote it if you are not wearing a belt and holster . . . and that is sticking in the front of your waistband.

There was a primitive sort of a dance hall on Eutah Street in El Paso forty years ago. Next night having exhausted the possibilities of the better part of town, a citizen of El Paso, a friend of Coley's and myself wandered along from one bar to another till we struck this place that was patronized by all the nations in the town. It was a one-story, adobe structure, plastered on the outside and in, it had the usual two narrow doors, there were a front and a back room with tightly packed earthen floors. The lighting was supplied by coal-oil lamps, one hanging in the center of the ceiling and some in brackets on the walls of each room. We had a drink at the bar, chatted with the proprietor who was dispensing drinks. He seemed friendly enough. He was acquainted with the fellow who was with us, although before this night, we had not the pleasure of knowing him. For some months, it seemed, he had been the guest of the State of Texas for having beaten another good citizen to it one evening. The beaten one became the "late lamented" in short order, and unfortunately for our companion he happened to have been a special friend of someone high up in the administration of the State's governmental machinery.

Those drinks were potent. Because I was a great hand, or I better say a great foot, for dancing any and everywhere in those days, I went into the back room where the dancing was going on. This room was a long one and it was lined with Mexican leather-bottomed chairs locally made. On these chairs were sitting a number of girls and some old women, most of them Mexicans. The girls were all properly chaperoned even though their experience in life was more than extensive; they were quiet and well-behaved. They danced well, and I started in dancing with one of them although her *dueña* remonstrated with her charge for doing so. She danced very well indeed. I realized that the few lamps at times looked to be a far greater number than I knew they actually were as the motion and rhythm gave the many drinks I had taken another chance to bite in.

The orchestra consisted of harpist, violinist, and *guitarrero* who played almost incessantly; finally they gave out and I took the young woman back to her seat and ordered something for us to drink, not forgetting the orchestra. We had only just begun to sip our refreshments when the *dueña* exclaimed:

"*Por Dios, aqui viene y esta borachо!*" By the Lord, here he comes and he is drunk!

I turned. At my right elbow was a tough-looking fellow who made a vile remark to the girl. Before I knew what was happening, he had slapped her a resounding whack and knocked her into someone's lap. He whirled toward me all set to go and I landed mine where it would do the most good from my point of view. He went over backwards, but he was quick and agile, he had swayed away from my blow and so was perfectly conscious. No sooner had he hit the floor than he drew a short-barreled Smith & Wesson. I jumped for him and reached for that so-called gun. It was a kind of disgrace in those times to tote that kind of gun and a greater one yet to be shot by the sort of man who would tote one. I was a shade too late to stop his firing it, but early enough to prevent anything serious happening.

He only succeeded in ruining my derby and singeing the hair on my temple.

At this stage of the proceedings the friend came and took the gun I handed him, while I yanked the man to his feet. I let him get his balance, and then operations were resumed. After a couple of punches or so, we carried him out to the street to cool him off, and put his gun back in his hand minus all but the one discharged shell. We started to walk off leav- him there, to renew our further diversions.

An enormous policeman hove in sight. The proprietor told him that the drunk on the pavement and some others had a fight outside there in the street; that he had never seen any of the combatants before; he turned to us for confirmation. Of course, we agreed with him, and said that he was undeniably correct. The cop was mad as sin at having a nice nap interrupted by what had turned out to be of no moment, helped the now-returning-to-consciousness slapper-of-women to his feet and took him off. . . .

XIX

Racial Characteristics

On the bright beautiful Sunday I spent in Chihuahua, I saw a queer manifestation of mob psychology. . . .

A bull-fight was planned and arranged for to feature a toreador once very renowned, but whose star was now on the wane because he was growing old. One cannot be a successful toreador after one is past young manhood. They go the way of most popular heroes. Gayety is irresistible to them, they live dangerously and with a feverish earnestness. All kinds of amusements of the most depleting character are indulged in by those physically courageous men who have the guts and dash and perfect synchronization to make them famous bull-fighters.

Although no man loves animals more than I, I must respect a skillful fighter of bulls for his aptitude, his expertness, and his courage. To walk into a bull-ring, to spread a pocket-handkerchief on the ground, to place both slipper-clad feet on that small piece of cloth and then to keep them there while the bull charges, moving the feet not one inch, but weaving the body itself in and out, back and forth exactly enough to make those needle-sharp horns barely miss the thighs or ribs, undoubtedly takes courage.

On this Sunday in Chihuahua there was an imemnse concourse of people at the *corrida*, it was a spacious *plaza de toros*. The band played a lively air, the *quadrilla* marched in. The occupant of the seat of honor, the Governor of the State, was duly saluted, the bulls dedicated to him.

There were six bulls. They were the very best, full of energy, wild and eager to fight before the *picadores* and the *banderierros* had goaded them to a frenzied fury. They were larger than medium, young, sleek. They put two or three of the *quadrilla* over the barrier several times, so cat-like in their swiftness, so full of the savage desire to kill.

The *matador* did his part well. He killed the first four bulls instantly and cleanly at the first thrust of the sword after playing them so damned close to him that his trousers, tight as they are, and his jacket both were ripped by the creature's horns. Naturally such furious bulls killed or disabled a good many of the poor, wretched horses before the fifth bull was let into the arena. The toreador presented himself, stood in front of the stand where the presiding dignitary sat, raised his cap aloft and asked permission to speak. This was granted him; he explained that with deepest apologies he was forced to ask indulgence during the remainder of the performance, he would be obliged to proceed without the *picadores.* He further elucidated that the contractor engaged to supply the horses had failed him and that he now had no more horses with which to continue.

At that, some individual (there is always some moron to do such asinities) started in to shriek objections. "They had paid their money for a fine show. They refused to be cheated!" Before the police could get among them, they had commenced a very pretty riot, throwing the light wooden chairs by the dozen violently into the ring. The bull-fighter retired. The police finally quelled the tumult. The show was declared off, the ring was left with broken chairs, backs and legs strewn about. The bleachers demanded that the old tame bull with the brass knobs on his horns should be turned now into the arena so that the amateur bull-fighters might have their sport, which is customary.

In came the big, slow, old fellow, he began chasing the amateurs here, there, everywhere. The temptation to prod him with the sharp ends of broken chairs was impossible to resist. They got the poor old bull so mad and in such an ungovernable rage that as he charged at a small group he reached one of the men and threw him into the air. The young fellow landed on top of his head. He was apparently unconscious, and his friends dragged him limply to one side of the ring and deposited him against the barrier. In half an hour this amusement was over, the old bull was allowed to retire. Then somebody remembered the young man who had proved the attraction of gravitation with his head. They went over to him as he lay motionless where they had dragged him. After repeated struggles to force him to stand on his feet, he was found to be quite dead, a perfect case of a broken neck. It was all ended now, except for the toreador, who was fined all of his gate receipts for non-compliance of contract.

Leaving the bull-fight, we strolled back towards town. Evidences of the Chihuahuenes' fondness for pleasures other than bull-fights, where no attention was paid for long to injuries or even to death, became conspicuous in more than one of the homes we passed where *guitarreros*, an occasional violin, now and then a piano, combined in making music for *bailes* in private homes; impromptu *bailes*, a group sitting about in the *sala*, while some young man with a guitar sang soft, harmonious love songs having to do with eyes, moonlight, soft roses, scented breezes, songs comparing the loved one to some particularly graceful warbler of the bird family. Young men with lovelorn expressions in their dark eyes and languishing postures leaned their shoulders against the bars of deep window seats, while whispering ever so tenderly of love and eternal devotion, *hacienda el oso*, making the bear. And why not? The transition from tragedy to mirth is nothing against a race . . . such men in my experience make good soldiers in the battle of life as well as on the field of battle.

Laughter-loving, courteous, mercurial, swift, tender, fierce, poetical, musical. . . .

Fatigue can be stalled off by sound, rhythmic sound. I cannot explain it, but if you could be with me for a while in a red-hot working where three or four sets of double-jackers were laboring, sweating in actual streams, and could hear the rhythm of their blows accompanied by chanting in unison, you would see for yourself. It is done unconsciously, the men fall into the tone accent almost automatically. When one or two have stopped for a minute or so to swab out the hole they are drilling it is a dramatic pause, there is a sensation of suspense until they start up once more breaking into harmonious chorus.

When the day or night shift would be waiting in groups at the head of the shaft or in some other convenient place for the *poblador* to designate the working to which they were assigned, they would begin a song or rather a sort of chant. In a moment fifteen or twenty of them would join in, and pretty soon all would be singing different parts, and sing them well. It was a pleasure to hear them and was also the cause of some speculation, because none of the men knew anything about music, none of them had ever been taught even the rudiments of harmony. It was just there . . . and their own ears and ideas of a concord of sweet sounds indicated to them what sounds to make in order to blend them into a tuneful whole. A discordant note would be immediately silenced by good-natured gibes such as "How sweetly the jackass brays!" "The young rooster learns to crow," "This axle requires grease," or some similar comparison with a primitive, unpleasant, or dissonant sound.

The Batopilense miners young and old adhered to a universal custom. There was a niche directly inside every tunnel entrance in which a cross was fixed and a constantly burning miners' lamp or a candle kept vigil. In front of this each man would stop and say a short prayer and cross himself. The older generation would frequently collect at the cross until there were some twenty or more congregated there, when they would chant a prayer or a few words of praise, cross themselves and go on into the mines to their work, a mile underground, to do their tasks cheerfully and happily. They were a grand aggregation of *hombres*, young and middle-aged. Quite often a set of two drill men would consist of father and son, and since under the tutelage of the older man the son became more and more expert, he would, in his turn, do a bit more than his share with the sledge in the holes they were drilling, to favor his father.

A *poblador* is a shift boss. Really he was much more than that. A *poblador* is born with the gift; he is the man who points the holes to be drilled, whether these be in a face, a rise, a stope, or a winze. The knowledge of how this should be done in order that each hole drilled and each powder charge exploded will give the greatest effect and break the most rock is an art that many never acquire, even though they be old hands at the game. The *poblador* ought to be a natural leader, one who has come up through the ranks, a strict disciplinarian, but with the right kind of discipline. He may have an assistant or more than one according to the number of the men on his shift. He has under him his peon bosses, who are in charge of those who are shovel men or laborers. The *operario* or drill man is of a higher class than the peon and is a skilled worker.

They liked their work. They were cheerful, contented, happy. They liked to sing, dance, laugh. The Batopilense was a great joker; his repartee was a delightful thing, at times possibly it was a foot or so over the line, but very rarely was it so in mixed company. He was innately respectful to older people and to his superiors. I have indeed known men of forty who would never think of smoking in the presence of their parents; very seldom was the word of the parents opposed; if it should become necessary to point out that the parents were in error, it was done in a very respectful manner. It may not be so now, for I write of an era that has gone into the limbo of forgotten things except by a few old codgers. . . .

The modern mining man little appreciates what the miners

achieved in order to get out their gold and silver a hundred, two hundred, two hundred and fifty years before we mined the native silver in that very old camp of Batopilas. The ores from these mines were, of course, world-famous. They were the only ones of the Western Hemisphere until the discovery of Nipissing, which produced native silver in the same amount and form with such varied massiveness and beauty of crystallization.

It was my good fortune to open up a very ancient mine that lay between two we were working. The company from which we had taken over in 1880 had no record of the existence of this ancient mine, nor had we, until my brother-in-law came across an ancient map of the San Miguel Group in the Department of Mines in the City of Mexico. He wrote us about his discovery, we investigated, and sure enough after a good deal of digging, we found what had once been a tunnel entrance. It was plugged up with heavy stulls of pine whose timbers were over fifteen inches in diameter. This tunnel was connected with a shaft to the surface about fifty feet farther up on the hillside, and below the tunnel the narrow little working had been taken down for three hundred feet by a system of short drifts and winzes. There were indications that very rich pockets of pretty good size had been taken out, since in places the working showed considerable magnitude.

We discovered what was of almost as much interest as the possibility of rich ground and these were the ladders and the butts of drill holes. The ladders were of the old native type, which means they were made from a pole fifteen or more inches in diameter with notches cut at regular intervals and at such an angle that when the pole was set against the side of the working, the bottom of the notch would be level and at least eight inches wide, where a man's foot could rest with perfect ease. Ladders of this type were no rarity but we had never encountered any made from timbers of such size as these. The ends of the drill holes—sometimes there was as much as six inches left at the bottom—were four or more inches in diameter. These had been drilled with a *pulseta* drill or jump bar, which is an iron bar six feet long, and on its end was welded a small piece of steel which was afterwards shaped by the blacksmith, usually the miner himself, into a chisel bit, and then tempered. With this tool held near its center with one hand and a little nearer the cutting end with the other the ancient miner would swing it forward against the rock; he would revolve it a little at each blow until with sufficient time and patience the hard rock would be bored to the required depth, scarcely ever more than three-quarters of a yard. When he had succeeded in doing this our determined miner of antique workings came upon his next great difficulty, this was home-made powder, naturally of a very inferior kind. The fuse followed this.

Before the days of paper, corn husks were used as the wrapper for fuses, or the inner bark of a certain tree. The fuses were made in this way: the powder was ground very fine, a small amount of this was rolled in the husk like a very thin cigarette, but no corn husk will average over eight inches in length at most in that country, it is safe to say that it would take six or seven pieces of husk to make a three-foot fuse. These pieces of husk must be stuck together with some sort of glue or paste, the gum had to be secured from a native bush or tree and fashioned into the glue for this purpose. Black powder must be tamped in order to yield any sort of rending force at all; therefore, after the powder charge was put in the hole and the fuse inserted, it had to be tamped. Very fine earth was used, sifted meticulously; this was tightened around the fragile home-made fuse in such a way that it would not be cut nor broken. Sometimes a thin rod of iron was put in the hole, the earth tamped around this and then it was later withdrawn carefully and the fuse inserted in the resulting hole.

When our ancient miner has at last been able to get this

far in his labor, we find that while a modern fuse would burn very slowly, thus giving him time to spit his shots and put himself in a safe spot, the home-made fuse flashes almost immediately to the other end when one end has been lighted, therefore, he would be blown to bits every time he lighted the charge or shot. Consequently he would take a piece of frayed rag (we are presuming the work was done before the days of cotton wicking), procure some beef tallow, boil it down and allow it to cool. When it had become quite hard, the frayed rag worked full of tallow was bent around the fuse; it was twisted tightly so it would stick up vertically and the end was pinched into as small a point as possible. He would now be all prepared and I think he would be saying his prayers. Cautiously, I am sure, he would light the very tip end of the fuse and LEAVE! The minute the tallow would feel the veriest breath of heat it would soften, the wick would fall to one side, and the fuse catching fire, an instant explosion would result.

I honestly feel that those old boys were entitled to all the silver they got.

Jesus Hernandez, the son of the old *Jefe Politico* of the same name, grew up with us boys and worked with us. The name Jesus is used in Latin-America as a given name as frequently as Thomas is in the United States. He was working with me the morning we first went into the ancient mine and to this day I can see his nauseated expression when we examined a strange pink growth of what appeared to be of a fungous nature on the foot wall in the little tunnel. We held our lamps close, the peculiar mass seemed to shiver, quiver, move. It did wiggle! What new discovery had we made of a scientific nature? It was a perfect conglomeration of very young bats, each about an inch or a wee bit more in length; there was such incredible number of them that they covered a space on the foot wall two feet by three. We shoveled them out onto the dump we had made in excavating the entrance to the little tunnel; later in the day a couple of old buzzards appeared to relish them immensely, who thought them, no doubt, tidbits of a rare sort—as indeed they were. Held by the tips of their wings the little creatures were as pretty a shade of pink as you would wish to see; they looked distinctly and precisely the way the devil was generally pictured back in the old days when we believed in devils and punishment after death for earthly sins.

Jesus Hernandez was a philosophic soul. On a certain Monday morning I was sent for to go to him in the hospital; he had been taken there the night before with an eye ruined and a terrific cut on the forehead. It seems that a rival (Jesus was a devil with the ladies) had waylaid him on his way home and had knocked him from his horse with a rock of ponderous size and inflexibility. The surgeon had told me that aside from the loss of his eye little or no damage was done. When I expressed my sympathy to Jesus and said how very sorry I was at having to tell him what the doctor had said, he replied:

"Ah, *mi amigo*, for all there is to see in this town, one eye *es muy suficiente*."

The sight of his remaining eye was evidently unimpaired for he chose one of the prettiest girls in town to make her his wife; it was at *carnaval* time, I think.

When you have a soft-boiled egg, scrambled eggs, an omelette, the shells of the eggs go their own way with a carelessness derived from the mental certainty that they are as useless as anything could be that makes up the contents of the garbage pail. If you have laying hens, these kind good females will eat them and return you a perfectly good fresh egg having renovated the original shell and made it do duty once more . . . but most egg-users don't raise chickens.

If you had been a citizen of Batopilas in the years that I lived there, and I believe even at this late date, the usefulness of the egg shell was only half completed when it appeared in the role of container. *Carnaval* time came regularly once a year in our camp, just as it does in New Orleans, Nice, and other places in the world, with the culminating international celebration at Coney Island in the sovereign State of New York. Among many other differences in the way these carnivals are celebrated is the use in Mexico of *cascarones.* A *cascaron* is an egg shell, about three-quarters of the entire shell. When the contents have been used, the top of the shell only having been cracked off, there was left a nice little container. These are accumulated during the year; if you are in the business of making *cascaronies* for *El Carnaval*, you engage all the neighbors' egg shells and agree to pay for them. You must lay in a supply also of many colored sheets of tissue paper, several of gold and silver tinsel, also colored dyes or paints. *And* a devil of a lot of patience!

When time for *El Carnaval* approaches, you stain or paint with water colors these egg shells in all sorts and varieties of tints. You cut the tissue paper and tinsel into such fine particles that they look, when mixed together, like a pile of many-colored grains of fine sand. Take half a teaspoonful of the paper dust thus made, put it into one of the stained egg shells; paste a little piece of the colored tissue paper over the open end of the one-time egg shell and then you have a full-fledged *cascaron.*

When you go to the carnival ball you carry your *cascarones* with you, when you are strolling or promenading around the ballroom greeting the guests, you crush one of them over a head allowing the shower of many-colored paper particles to fall on the hair of the lady thus favored. She thanks you prettily, because it is a sign of respect and indeed of affection; I have seen an especially pretty girl so covered with this dust at the end of the ball that you could barely tell the true color of her hair. It is a novel kind of confetti, really.

Not only while at a dance but during calls or at any other opportunity in carnival time this is the thing to do, crush the *cascaron* in your hand as you hold it over the head of the favored one and release the contents.

Like the ichthyosaurus, chaperones have gone out of fashion, but they still roam in Batopilas I am certain, so be sure to devote several *cascarones* to the chaperones. It does not permanently endanger their vision, but it has a strong tendency to make them see things as through a softer and less Grundy-ish light.

Now, the Mexican is usually a careful person concerning the waste of foodstuff; the beloved and good *tortilla* is made from corn, the wheat flour *tortilla,* however, is not despised, and bread and rolls are also used by the more cosmopolitan of the population. You may say: "What has that got to do with *El Carnaval?*" There is always a big battle on one chosen day during *El Carnaval* between the Abajeños and the Arribeños, the people who live in the down-river part of the town and those who live in the up-river part. The weapons, or better, the ammunition, are flour, white wheat flour; and you can put anyone out of the fight if you are quicker than he may be by filling his eyes, mouth, and nose with handfuls of well-thrown flour. It is very unpleasant, for your hair becomes full of it, it blinds you temporarily, chokes you more than considerably; unless you have the good sense to wear white cotton or linen clothes you look damned funny when you have done your bit in this battle. Once or twice during our years in Batopilas there was a case which caused real suffering, because some simpleton thought to get even with an enemy by substituting lime for flour; if he were identified he got it back with compound interest, be sure of that.

XX

Accidents

Batopilas was a hard-rock camp. The nature of the ore deposits came at first glance apparently in more or less irregularly arranged *bonanzas* or rich deposits varying in size from ten thousand to a million or more dollars. When drifting and sinking in the effort to discover these rich deposits many holes must be drilled and fired. Dynamite is, of course, the effective explosive. Sometimes we used forty per cent but usually it was sixty per cent.

When a Batopilense celebrates he likes a noisy celebration. During the visit of the Governor to our camp more than one case of dynamite was fired in salutes. It is a very efficient way to fire salutes if you keep sober and are not heedless in the handling. However, if you are not careful you are more than liable to cause as well as to suffer trouble. It was not a rare sight in Batopilas to see a man now and then whose right hand was missing—the outcome of too much enthusiasm in combination with liquor, dynamite, and a Saints' Day celebration.

It is done in this way: You procure a stick of powder, cut it into four pieces; get four short lengths of fuse, about six or eight inches, slip caps on them, crimp the caps. Generally this is done by putting the cap in your mouth and crimping the edge by biting it. Next make a small hole in the end of the dynamite, insert and spit the fuses, and at the exactly proper time you cast it into the air, or some place far enough from those present to avoid endangering them. It is a well-known fact, according to the very best Prohibition authority, that among other devastating effects of alcohol, it distorts the judgment. Therefore, if after lighting the fuse you start in making a speech or otherwise allow yourself to be diverted from the actual situation in hand, when the resounding explosion is over you will find that your hand has also disappeared with the smoke and the noise.

If you want to make a cleaner job, you can use a third or even a half stick of dynamite. But this is not recommended for those not possessed of an above-the-average strong and robust constitution . . . better ask the surgeon.

Dynamite is a strange substance, kind of like a woman in some ways, for they occasionally do not function according to the rules and customs which are supposed irrevocably to govern life. I remember one time a man was sent to the powder magazine which was in a distant and abandoned drift. He was careless with his light and somehow three cans of black powder became ignited and exploded. Within twenty feet of the powder cans were three cases of forty per cent dynamite. The wooden cases were thrown about and scorched, but they did not explode. Not possible? Well, maybe so. But it happened. I was one of the first men to reach the place, and in a thick and nauseating smoke I helped to pick up and to carry out the dead and well-broiled miner who was responsible for the catastrophe. When things had cooled off and the atmosphere had cleared somewhat I examined the spot and the facts were as I stated them.

I have seen a mule—you see I cannot keep away from mules—loaded with four cases of dynamite roll down a steep hillside for two or three hundred feet, scattering the dynamite from Hell to breakfast, and nothing happened. Once, though, I saw a mule do it, and after rolling he dropped vertically a hundred feet. There was a tremendously loud noise . . . that is all we recovered.

When the Chihuahua & Pacific Railroad was built out to La Junta our stage drive was cut down to sixty-five miles. One morning I was waiting there for the train to come so we could load our bullion and go on to Chihuahua. I began chatting with the agent and others. We were leaning against a pile of cases of dynamite stacked on the platform. As the train pulled into the station, we noticed there was an extra passenger coach. This train was of such modest proportions normally that an extra car, especially a Pullman, was an object of particular regard.

"Oh, yes," said the agent, to our inquiry, "that's right, certainly it is a Pullman. It is a special car. He is making a trip over all Mexican railroads."

The bullion was stacked at one end of the platform of the station where the express cars of the train would be. The dynamite cases were at the other end and the wagons were preparing to load it. The observation platform of the private car stopped immediately where we were doing this. A gentleman with a party of friends was sitting on the observation platform and as I glanced up I saw the expression on his face blaze into consternation and fear, he half rose from his seat and gasped out:

"Look out! My God! Look out! Pull the train out. You damned fools, throw away those cigars, you are going to blow us all up."

With this he rushed to the other side of the observation platform and leaped over the rail. He was clumsy, fell, and rolled down the slight embankment of the road. It flashed upon me that the ends of the cases of dynamite had that word stenciled upon them in large letters and that, therefore, this was the cause of all the activity. We ran around the end of the car and picked him up; we had forcibly to restrain him from a sprint across the plains. After some brushing off and other soothing efforts, I prevailed upon him to calm down. I explained to him that there was no danger to the other members of his party who were still seated on the observation platform of the Pullman, and who were regarding him with expressions of combined amusement and timidity. I impressed upon him the fact that the railroad company itself would never have unloaded the dynamite at that spot if there had been any peril in it, any danger of blowing up one of the two trains which it owned. Moreover, the company was entirely aware that the private car with many important persons was attached to that particular train. In fact, that no railroad company could afford to take such a risk. Hence, there was no actual jeopardy.

"But, by the Lord Harry, sir, you were smoking!"

"Well, then we will not smoke any more until we have gone three or four miles up the track."

Eventually I succeeded in getting him back on the train. His poor wife, to whom I was introduced, was half way between tears and laughter. The efforts of the others to refrain from roars of amusement plainly showed him to be a person of great importance, but not of immense sense of humor.

Just like anything else, continually handling dynamite becomes second nature, and you automatically use caution and circumspection in its management. There are men who are such experts in putting and shooting dynamite that, if you had a big boulder in the middle of your drawing-room they could crack it with dynamite and not soil your wall paper. It is all in the knowing how.

One day we were examining an old working, a big stope, to see about putting in some timbers. There were three of us. We felt a shower of small stones and we jumped aside. This shower is generally a forerunner of a larger piece of rock becoming detached and falling. Two of us barely got clear. We expected the other one to be squashed flat as a pancake. In a minute we heard him groan, and we spoke to him. When we could get to him to examine him we found he had a bump on his head as big as an egg, but he had escaped further injury. Three big boulders were holding the slab of fallen rock and he had been able to throw himself prone between them. We had severe labor in getting the slab from off him and extricating him. We were obliged to shoot one of the slabs with dynamite and that took damned careful calculation as to exactly how much dynamite to use in order to crack only so much from the slab and no more, consequently not to let it completely down on him and thus crush him to death after all.

Naturally only a part of the time does one come out unscathed from dynamite accidents. Then it is a nasty business carrying mangled live and dead men on your back up or down several hundred feet of ladders. They will bleed all over you and the odor of human blood has always been repugnant to me. However, if the mine management and discipline are as they ought to be, you will find that at least ninety per cent of the accidents are due to the man's own carelessness and because he has not complied with the laws laid down for his own protection.

One night I was awakened by the night watchman at San Miguel Mine with the word that there was someone hurt in the mines. I jumped into my pants and shoes, ran to the gates, the watchman handed me my lighted lamp as I passed him. I opened both the outer and the inner gates. The messenger from the bottom of the mine told me that so-and-so had been killed in such-and-such a winze by having three charges of dynamite fulminate while he was still at the bottom of the winze.

In I hurried to the mouth of the hoisting shaft. I was informed that the injured man was being carried up the ladder-way on the back of a peon. They thought it best to bring him out that way. It did not take long to do it; as I have told you, those peons could put as much as a hundred and seventy-five pounds on their backs and climb ladders with very little effort. I lit out down the ladder-way. We met the procession and found the "dead man" slightly cut up about the side and back, but quite alive and cheerful, smoking a cigar!

We sent him to the hospital and he was out in a week. Of course, by all the rules he ought to have been dead. Three one-yard shots had gone off in the bottom of the shaft six by ten feet and he was right there in the bottom of the shaft with it.

The reason for his being there was his own laziness. He ought to have secured a new rope from the boss with which to climb the twelve feet to the end of the second ladder. The lower ladder is always removed when the round of shots is to be fired. The hand rope he used was old and frayed, this broke with him and he fell back into the winze. He was slightly stunned, therefore unable to call to his partner on the level above him. This worthy had also disobeyed orders by not remaining at the mouth of the hole until his companion had come out. The injured man was a very wise miner. He knew that the shots were pointed to throw against the foot wall. He scrouged himself into as small a ball as posible, well down in the corner opposite to the shots, and said his prayers.

The charge exploded, breaking three tons of rock, of which only about half a ton fell back onto him. After he had regained his breath and was pulled out from the stuff, he asked for a cigar; this was given him. He remarked, "*Dios es grande!*"

One of our employees had married a Hot Country girl and went there to bring his family back to camp. On his return he told me about a colonization scheme that was going on at Topolobampo on the coast of the Gulf of California, not far from El Fuerte. This port of Topolobampo is entirely landlocked, very deep, an uncommonly good natural harbor. The colony of Americans there was unsuccessful and we got two or three fine young fellows from that place when they had to leave there to seek employment elsewhere. I had a sad experience with one of these boys. He worked for us a number of years in charge of a shift on the Big Mill. But he grew tired of it and wanted very much to get into the actual mining end of the game, so *El Patron Grande* sent him up to me at San Miguel Mine.

For a few days I took him about with me to familiarize

him with the mine, and then one day I sent him up to a level eighty feet above tunnel level where we had a working in light silver.

I went about my own affairs; going into the shaft I jumped the skip and was dropped down to the fourteenth level where we also had silver. I had been there scarcely twenty minutes when a peon came running into the working where we were getting ready to fire a round of shots, with the word that "Don Gustavo has fallen down the *chorrera*." This is the chute through which ore is thrown to the tunnel level.

We ran to the skip and were hauled as rapidly as might be to the tunnel level. As I went by the chutes at the hoist, I called off all the car men to follow me and sprang out towards the main tunnel. When I got there to the chute I had another demonstration that *Dios es grande*. We carefully removed the gate of the chute; I had two of my car men to hold back the ore with their shovels, and I climbed in only to find Gus lying on top of what was the very last few carloads of ore the chute held.

There were still several carloads of rock caught up above me and it would require but the slightest jar to start it sliding down on us. I got my hands under the boy's arms, strongly braced myself so that we would not slide down with the rock as a sufficient amount was drawn off to enable me to pass the now unconscious Gus, very bloody and breathing heavily, out to the waiting men, feet first. This we managed to do, and we rushed him out of the mine; there the doctor was waiting for us with a stretcher. He gave Gus a hurried examination and had him taken to the hospital. Gus did not remember a single thing about any of it afterwards. The only words he uttered were, "For God's sake get me out of here!" when he heard my voice.

He had rolled eighty feet down the chute. He must have been walking in a sort of dream ever to have stepped into it for it was a hard feat to accomplish even if you had set you mind on doing it. A lucky thing it was for Gus that a loaded car was just about to be dumped as he got there, for the car man had paused to let Gus pass before dumping the ore and he had been a witness of the fall. If it had not taken place in this way Gus would have been covered by tons of rock, because we would not have drawn off that chute until the next day. There would certainly have been the devil's own time to find him and to get out his body, or what was left of it. As it was, Gus stayed in hospital only about two weeks.

But at my special request he was not returned to the mine. I did not consider him noticeably qualified for this kind of work.

XXI

Fighting Cocks in Mexico

In Mexico a cock-fight is a legitimate pastime and is conducted openly and aboveboard. The atmosphere surrounding it is more like that which used to exist in Ireland and even in certain sections of the United States before the "purists" began to clean things up.

The training and heeling of cocks for fighting is an art that few understand. It was really an art, the man who was an expert in tying a knife on a chicken was one who was in high demand. The *ammarrador* had to know his business. There was as much care in the training of cocks as there is in training for a prizefight in the United States. The chicken is a courageous bird and fights for his life. One of them is always killed and very frequently both combatants die in the ring. It is nothing like so ignominious a death as to have his neck wrung to serve for Sunday dinner. When you possess a chicken that has weathered three or more fights you have a marvel, few reach this number of fights. He is thus proven unusually good at fighting and is put out for breeding purposes, the resulting chickens for which he is responsible are birds of value.

In Mexico the gaff as we know it is not used. The weapon is a double-edged, minute saber as sharp as a razor both as to edges and point. I have seen a doctor perform a neat operation with one when his regular tools were not available. They vary in length from two inches to three and are shaped and balanced and of appropriate weights to suit the particular cock, his size, weight, general conformation, and style of fighting. The natural spur is cut off leaving a stub less than one half inch long. The *botanilla,* which is the pad with a hole in it to fit over the spur-stub, is made of the finest chamois or buckskin. Around the hole two or three thicknesses are added so that it makes a pad three-quarters of an inch square with the hole exactly in the middle. The ends of one thickness are long enough to overlap when placed around the leg of the chicken. It is made with the utmost care and the weight is very slight.

On the end of a fine twisted silk or cotton cord is a loop which passes over the spur-stub, is wrapped twice about the center, once or twice between the spur-stub and the edge of the *botanilla* above, and the same below the spur-stub.

The knife, the *navaja*, is now produced having been skillfully chosen from a selection of twenty-five or more, which every owner keeps to fit the requirements of this especial cock. What would be the hilt, if this knife were a saber, is at right angles to the blade and projects like the letter "U," the saber going from the base of the "U." This is slipped up over the spur-stub and settled firmly and evenly on the *botanilla*, then made fast by wraps around the leg and *botanilla*. Finally after it has been pointed with exactitude it is secured tightly by the last few wraps of cord.

A leather scabbard is fitted over the *navaja* in order that the bird may not cut himself or the person who looses him into the ring. This tiny scabbard remains in place until just before the cock is released; after he has been released no one except the judge can touch him until the fight is either won or lost. The rules are very strict and are rigidly enforced. The judge must also be an expert. Those who release the cock, the trainer, the *ammarrador*, must know their business if they are to win any fights.

Now, a game chicken must be as tame as a pet dog. During the training he is staked by an especially arranged leash, and this leash is always fastened to the leg which takes the knife. The trainer gives him his food and exercises him, allowing him at stated intervals a "go" at another rooster about twice a week for a short tussle. In the meantime, he exercises the fowl in dodging and turning and striking by holding another chicken in his hands and playing this other chicken with him, thus teaching him to jump and to strike at different angles and distances from the ground.

I have had a game chicken spring up to strike another one that I was holding as high as my shoulder. The strength in the blow of a trained cock is sufficiently powerful to leave your knuckles numb if he hits your hand. If he is tame and accustomed to being handled he will not be startled nor made nervous when he is put in the ring surrounded by the tumult of noisy, shouting onlookers. You always match your chickens according to weight, unless they weigh over six pounds. From that weight upward they are what is called *tapados*, or free for all comers.

All this may give you an idea of what it takes to keep a string of fighting gamecocks in training. It is an exact and a nice science.

There is no bird more beautiful than a chicken when ready for the cockpit; that is, the way they train him in Mexico. He is always in complete and full feather, not one of them is out of place nor missing. You may not know it, but the big stiff feathers in the tail of a rooster keep him from being knocked over backwards. He goes backwards spreading his tail, and each feather acts as a part of a fan-shaped prop.

Game chickens once were the cause of my spending forty-eight hours *incommunicado* in the public school building in Batopilas, which was used as a substitute for the regular jail for two reasons. First, the jail was too full. Second, it was lousy and ill-kept and the old *Jefe* did not want anyone whose unfavorable comments would receive attention in the office of the Governor in Chihuahua to have personal and undeniable experience of conditions.

This *Jefe* was not a native of Batopilas, but had been sent to our camp for some political debt owed him from the City of Mexico. We will call him Don Bombardioso Cogñaqueño because he was bombastic and he was too much addicted to the sipping of cognac. He was tall, about sixty years old, wore a long beard, which with the hair on his head he kept dyed a glossy black, because he was a dog with the ladies . . . or thought he was.

Don Bombardioso was a devoté of the cockpit, and he believed that he knew more about the sport of chicken fighting than anyone else in the world. He challenged my kid brother Conness and me to hold a main, a *pelea de gallo*, on the fifth day of May. There were to be five fights; four matched for weight, and one *tapado*, making five in all. Five hundred dollars for the winner of the main, the majority of five.

This *Jefe* did not like us. He had it in for all Americans. He brought this dislike with him to our town. We had always been perfectly respectful to him, but he was not partial to the conditions in Batopilas where all our workmen were contented, their respect for us as individuals and for the organization which had made their lives much easier than was usual in a mining camp, being so very apparent. He also realized that he was actually second in command to my father in the town and in the minds of the majority of the population; also he himself was unpopular. As the Batopilense reasoned, was he not an outsider? Did we not have citizens every bit as capable as he to be *Jefe?* He did his best to make trouble for the company and was instantaneously squelched in his efforts by the Governor of the State. All of this, added to his probable defeat on the fifth of May, which was a day of rejoicing, a national holiday, made the cockpit crowded to overflowing, and was almost more than he could endure.

The great day arrived. Old Don Romulo Rocha was the *asentista*, the judge, at the chicken fight. I was appointed to release the cocks belonging to Conness and me and the *Jefe* was to do the same for his. All went well until the very last fight. We had won the first four, and Don Bombardioso was boiling. His temperature reached a height causing him to try twice to interfere and pick up his bird when it became entangled with the other cock in the pit. This, of course, is strictly against the rules; once the cocks are loosed in the pit, no one with the single exception of the *asentista* being allowed under any circumstances to touch them.

The first time he tried this, I was polite, but firmly remonstrated with him. The second time he sprang forward and stooped to grab hold of his bird, I pushed his shoulder. He sat down hurriedly and in no dignified manner. The watching crowd howled its appreciation. A bystander aided him to his feet. He made a mighty struggle to welch on his contract, and we agreed—"If he thought that anyone claiming to be a *caballero* could do such a thing."

Confidential orders were given to the Chief of Police to see that one of the policemen should do something that would "compel one of those *gringos*" to strike him, and then the fun would begin.

Old Macario, the Chief of Police, had known us since we were small boys. He had a friend in us when he needed funds, or hospitalization for himself or family, or for any policeman or prisoner in jail who got sick or was injured; he had been hard ridden by the new martinet of a *Jefe*. What, therefore, was more natural than that he should quietly caution us what we had to expect!

But it all supposedly blew over, and we thought everything had been forgotten, since we obeyed the warning Macario had given us—to keep away from the town for a while.

Two months had passed, and we believed the whole affair lost in oblivion. Bill Kirby-Smith and I were returning kind of late from down river one afternoon where we had been inspecting some prospects that were being opened up. We came through the town and passed the home of a friend. An impromptu *baile* was in progress; our friend, who was a wild youngster, came hurrying out when he saw us and insisted that we come in. He urged us so eagerly that we agreed, but not being dressed for dancing we only went into the refreshment room to have some beer. Within half an hour we said good-bye. As was customary, I let out a yell in the street and a "*Viva, Sarachito!*" the name of the young friend whose hospitality we had enjoyed.

Then I walked over to my horse, took my spurs from my *mozo*, and stooped to adjust them to my feet. At that precise moment, a policeman ran up to me and using obscene language reached for his gun. Of course, I hit him. I knocked him down and extended my hand for my own gun which my *mozo* was holding. I had left it with him, because it is exceedingly bad form to enter the home of a friend wearing a gun or spurs. The Chief of Police hurried to the scene. He ordered the cop away and told us to go home and "Be good, please." Bill Kirby-Smith had leaped across from where he was about to mount his horse, gun in hand. I had some difficulty in convincing him that the best thing to do was for us to obey Macario, that he was our very good friend and that we really ought to follow his advice.

Alex was then General Manager, and next afternoon he sent for Bill and me. We went down to the Hacienda San Miguel and he showed us a letter from the *Jefe*, in which he stated that we had been guilty of a serious offense and were to come to his office in the town to pay a fine of one hundred dollars each. We went to the *Jefatura*, which was on the second floor of the building, and there the old rascal proceeded to read me a lecture. I told him:

"I am here at the request of the *Jefe* to pay a fine for having struck a policeman who without cause grossly insulted me. But, with the *Jefe's* permission, I am not here to be the recipient of a lecture on good behavior from the *Jefe*."

He further told us that it was not a matter of a mere fine. (Although he had accepted the one hundred dollars from each of us and had given us each a receipt for it.) He informed us that we were to be incarcerated until a decision should be reached as to our ultimate sentence by himself, the *Jefe Politico.* I knew that all this was the bunk, of course. But I saw, and Bill did too, that the only exit from this upstairs room was filled with *pelados,* armed troops, and their bayonets were fixed. I agreed to be incarcerated. I didn't want too much trouble nor to make any move to place me in a position where he could say with truth that I had resisted an officer in the performance of his duty. We were escorted to the public school. It was vacation time and the building was empty. This was a nice structure adjoining the jail. Passing my *mozo*, who had been holding our horses, en route to the "jail," I told him to tell my brother that Bill and I were *incommunicado.*

Almost at once we were supplied with comfortable cots and bedding by some of our sympathetic friends in the town and as soon as supper time came we were served, from the same source, with more delicacies than we could eat. The jailer handed us nice little notes of joking condolence. So much for the lighter side of the affair.

On his way to the Hacienda San Miguel, my *mozo* spread the news. He met some of the day shift from San Miguel Mine and the Porfirio Diaz Tunnel, and told them. They ran back to the mine and soon from both places they had organized a hundred or more miners under the command of the two head men, and had started for the town to take the damned jail, to clean up the police, and to free us from durance vile.

But not so vile. Bill and I were in separate rooms, to be sure, but at that moment we were engaged in consuming all sorts of good things to eat.

Fortunately my brother Alex and my cousin Dr. Merchant were in the act of mounting their horses to go to the telegraph office to wire the true facts to the Governor of the State at Chihuahua. They hastily crossed the river and were able to intercept the army of angry miners and to explain to them that their efforts would cause no end of trouble and difficulties to all concerned. Alex received word from an influential citizen that the entire business was a frame-up, that he could accomplish nothing locally, for the *Jefe* was saturated with cognac and was, besides, in a very ugly humor, refusing to listen to anyone. My brother was further informed that a garbled account of our conduct had been sent to the Governor the night before. This report stated that I had fired six times at close range upon the Chief of Police, but fortunately had missed him.

Good old Macario had refused to confirm this by a sworn statement at the request of the *Jefe*. First, said Macario, it was not true; secondly, anyone, although not a witness of the occurrence, would know it was false; and thirdly, there were many persons present who would swear that my gun and belt were still held by my *mozo* when it all took place.

Alex sent a true statement to the Governor. This gentleman had telegraphed the contents of the *Jefe's* telegram to one of the leading citizens of the town with the request that it should be confirmed or rejected. A prompt rejection was wired to the Governor with the additional clause that the claims were false on the face of them, because the sender of the message knew that it would be an impossibility, if I had fired six times at close range at any man, for that man still to be walking about the streets uninjured, as the Chief of Police was doing at that precise instant. The Chief was also willing to "deny the whole thing as being a fabrication and that under no conditions could he see me, whom he had known since I was *muy pequeño*, do anything unkind to him . . . unless it were to make him drink more liquor on occasions than might be good for him, but that he would leave it to the judgment of the *Señor Gubernador* if that was an unkindness."

XXII

Thieving Silver

About the year 1898 a renegade labor-union cast-off came to Batopilas and instigated a campaign. We said nothing but left it to the men. For a few days about ten per cent of the miners stayed away from work; suddenly they all came back. The older men knew that nowhere in that part of the world was the worker so well taken care of, and they went to the Chief of Police, took council with him, and abode by his advice. He was their friend and had been a former companion.

Of course, the Chief could not take action officially, but he intimated to the old miners that they ought to call upon the trouble-maker in a body and suggest to him that the miners were entirely satisfied in their work and with their pay; that they had no intention of supporting a lazy good-for-nothing; that they certainly were not fools and they realized that the company had given them fair treatment. The Chief further instructed them to let the cast-off union man know that their Patron was their friend; that he provided them with a good hospital; that he let them have money when it was needed in emergencies; that he charged them no interest, allowing them a long time in which to repay it; that unless this newcomer was in a position to duplicate or to improve these conditions he would do better to go away. This he did. That was the last as well as the first we ever heard of him or any other labor agitator. It was a joke for a long time among our men.

"To think," they used to say, "the poor fool believed that the Batopilense was such an *inocente* as to be taken in by such a *tonto*."

Although the men were faithful, in every group of employees there are a certain proportion who will take the easiest way. . . .

The ores from our mines were what is commonly known as high grade; therefore much watchfulness had to be exercised to prevent them as well as the silver itself, while in the various processes of reduction from the crude ore state to the refined bullion, from being filched. We had a group of trusted employees, native and foreign, who kept themselves on the thinnest edge of readiness to combat the ingenuity of those who hoped to add to their wages (in a few cases they succeeded in doing so) by stealing small amounts now and then and carrying them out of the mines on their persons. Some of the methods are worth relating—there is no knowing when you may be reduced to working in a silver or a gold mine, and with this knowledge of how to secrete the precious metal you might make enough on the side to cover your monthly installments on the radio or the electric ice-box.

Say, for instance, you wear your hair quite long, and while you are turning the drill steel and it is cutting through rich native silver ore, you clean your hands to rid them of the mud caused by the cutting of the drill-bit moistened by the water you squirt from your mouth into the hole; if you succeed in cleaning your hands by wiping them on your head the long hair will get full of mud. When you go home, and wash your hair in a basin, you will find that there remain in the basin fine particles of silver: after you become expert at it you will be able to run it up to as much as fifty cents worth a day. If you are diligent this will increase your weekly income about three dollars.

Suppose, on the other hand, you are wearing the native sandal made of heavy sole leather. If you exercise patience and skill you can insert the cutting edge of your sheath knife into the edge of the leather, preferably on the inner side under the instep. Gradually as you work this edge back and forth to split the sole leather, you will have an opening not much wider than the blade of your knife; thus you make a pocket into which you can put an ounce of native silver that you have secured by pounding up some of the rich ores and blowing the powdered rock away, leaving the heavier silver.

Another way is to forego your dinner. Never mind if you are hungry; if you will take your dinner of hash or *frijoles* into the mines in a porridge bowl with corn *tortillas* laid on top to prevent the contents from spilling, the whole thing wrapped in a clean napkin, and can go without eating it after working hard all day, you can put the filched silver in the bottom of your bowl. You can replace the *tortillas* on top of the hash or *frijoles;* look sick as you reach the mouth of the tunnel to be searched; tell the gate keeper and searcher that you did not eat your dinner because you had a *dolor muy fuerte* in your stomach, and you may get away with it. But you will stand a slim chance indeed if old Don Romulo Rocha or Ramon is on the job.

There are other very unique methods, but the ones I have told you about will be plenty to start on. Oh, yes, there is another trick you might attempt. However, it is scarcely worth while because you would have to establish a reputation as a *macouchic* smoker. This is a very inferior tobacco. Learning to smoke it is in itself an ordeal that badly shakes the system of the uninitiated. Tie up in the corner of your bandanna handkerchief a handful of *macouchic* and mix the fine silver with the *macouchic* before you come from the mine. But Don Romulo is so very expert that he may smile subtly and say:

"What heavy tobacco that is! Let me look at it; I am, as you know, an expert on tobaccos."

He will do the same thing with your cigars, so do not waste any time hollowing them out and charging them with silver. I have watched that fine old fellow, Don Romulo Rocha, search hundreds of men and I honestly believe that he intuitively knew, when they stepped up to be searched, just who

Sampling the vein.

Searching at the gate of Todos Santos.

Bullion Shipment—About 300,000 Ounces Fine Silver. Each bar weighing from 60 to 65 pounds.

did and just who did not have contraband silver on him. Many of the mines were superstitious about Don Romulo. When he had been tipped off by the mine boss—which used to happen rather often—that so-and-so would probably make an effort to pass him with some stolen silver, he would wave the man back into the tunnel a few feet until he had finished with the others of the group. Then he would call the culprit and say:

"I am surprised and hurt that you should have attempted to do this. Give me the silver. Appreciate the fact that I do not expose you to the rest of the men as a rascal and a thief."

That old gentleman was a student of psychology. He could have given cards and spades to some of the persons who write books on the subject.

But taking all things into consideration and the temptations these miners were subjected to, there were comparatively few who tried to steal. As a matter of fact, at first they did not look at it in that light. "Has not *Dios* put it there to be taken from the earth and to be used by mankind? Did not my own labor actually bring it forth?"

An old woman, Juana, put one over on me. I never did live it down because I was supposed to be tolerably wise about such things. She showed up at the mine one day to draw her pension (her son had been killed at the mine some time before) with a hunk of ore, a good-sized piece and very rich in black silver, sulphide of silver. Juana would not tell me from what place this ore had come, but said that, for a reward, she would show me the place.

"Well, Juana, you know what the rules are about that; if this proves as you say you will be well-fixed for the rest of your life."

I settled matters so I could go with her the next day, Saturday and have two men accompany us with an oufit to remain for two days, to shoot a few rounds of dynamite in the vein this rich ore was supposed to have come from. Very early in the morning I sent a *mozo* with a saddle-mule for the old woman, dispatched my men with the pack-mule, telling them to go ahead and that I would catch up with them on the trail. This I did after putting in the shift on Saturday morning and signing the payroll. The superintendent's O.K. was necessary for the money to be delivered to the paymaster. Off I went in high expectation.

About twelve miles up the river I overtook the cavalcade; we turned off, and continued upward into a very bad trail that led to God knows where. I noticed that Juana had a good-looking young Indian girl with her, she too was mounted on a mule. After we had arrived at the place where Juana insisted that the vein was located, I knew that she had misrepresented things to me, and that the only object in the expedition was the twenty-five-dollar store credit I had arranged for her. The good-looking young Indian girl had been brought along as a sacrificial offering to appease my expected wrath.

It was growing dark when we reached this place, but I emphatically affirm that any old woman who had endured the ten hours' ride on a mule over those trails as Juana had done for twenty-five dollars store credit, did not deserve too rough handling.

At break of day I had a look at the supposed vein; there was a seam of decomposed porphyry; I sampled it and traced it for some distance. There was no possible hope that the rich sample she had used as a bait could have appeared in the formation in that section, and I gave Juana a mild reproof. To the Indian girl I gave five dollars so that her hard journey would not prove a total loss, warned Juana that if she took the five dollars away from the girl I would have her put in jail, saddled my beast and lit out for camp. I got back to the mine in time to have a shower, eat my supper, and get to town in time not to miss the circus showing there on Sunday night, which was a rare event in our camp.

Being well-lubricated when the evening was a little older,

we were having *menudo* in Manuela's and I told our crowd, in an unguarded moment, about my fruitless quest into the mountains. I had to buy wine and more wine and then some more wine! Old Juana cost me more than a hundred dollars.

I sent word for Juana the following week to come and tell me where she had got that piece of ore. I was informed that Juana could not come to see me then, that she was getting married. Thus the necessity was explained for my contribution taking the form of an order on the company store. Further inquiry revealed that the man she was marrying was a powerful Mayo Indian, who was known to be infernally lazy but more than commonly attractive to the ladies . . . so much so indeed that he could afford to indulge his talent for being the most indolent devil in camp. I knew now who had done the thinking-out of the plan, and I sent a note to my friend the Judge.

Bright and early one morning up comes the old girl accompanied by her former lover, now her husband. He was all rigged out in new store clothes. I knew that unless I landed him now, both Juana and I would be the losers. I took the fellow aside and told him: "You *sin verguenza,* what did the judge tell you yesterday?"

"He told me, *Patroncito,* that if I did not pay that store order he would put me in jail."

"Right," said I. "Get your good clothes off. Go into the mines on the day shift. I have already told Don Manuel where to put you to work, and that work will be good for your soul. Old Juana may support you by taking in washing after you have paid off that twenty-five dollars, but until that time you work on Number eighteen level where it's hotter than the hinges of Hell . . . or you go to jail."

"I will go to work, *Patron,*" agreed the Mayo.

I learned that the Mayo had picked up that ore sample in the town dump and had cracked off the worn edges with a hammer so that it would look like a comparatively newly broken piece of ore. Further investigation revealed that Don Ignacio Ramirez had received this sample from a friend many years before while he was in Chihuahua, and the Lord only knows where it came from in the first place. It had landed on the town dump after a house-cleaning at Don Ignacio's. All these investigations and things took time and my best Secret Service methods.

I got the dope about the sample's reaching the dump from Don Ignacio's lovely daughter while sitting out a dance one night in their garden, where the scent of the orange blossoms and the light of the moon tended to soften the sting of what that wretched piece of ore had cost me.

The Batopilense is a great *chalan* and *pullista,* which being interpreted means a great chatterer and "ragger." But this ragging was a sort of play on words something like punning with, as a rule, a risqué tone and very witty. So that condemned piece of ore cost me much more than a couple of hundred dollars, what with the good-humored raillery and banter which grew less as time went on but which never wholly ceased.

XXIII

Comestibles

There was a custom in our time in Mexico, I hope it yet prevails, of serving *menudo.* This is a tripe stew accompanied with a red sauce of *chili colorado,* very hot. It took place mostly from one in the morning on until daylight, and certainly was a life-saver for those who were overalcoholized. It surely straightened you out. All the *fonderas* specialized in this. There was one old woman, Doña Anselma, a vigorous soul, whose *fonda* was noted for a *menudo* that was particularly efficacious and palatable. Her long table would seat as many as twelve and on Saturday and Sunday nights it was always working to capacity.

Rafael Diaz was a tall young miner, very pleasant in appearance. With some others he came into Doña Anselma's one night; there were two big *ollas* of *menudo* simmering on the hot coals in the corner. Rafael was observed to approach one of these while the old woman's back was turned. This Rafael loved a practical joke, and being well-stimulated added to this rather questionable taste in amusement. Coming along the street on his way to the *fonda,* he had been a witness to the fact that although a cat may be quick, the hind feet of a mule, at times at any rate, are even more alert. A miserable and unfortunate cat, chased by an equally wretched dog, had tried to escape by climbing the hind legs of a mule which was peacefully sleeping in front of the house where its owner was disporting himself with wine. . . .

Now, the hind legs of a sleeping mule are something to be treated with respect and consideration. If you must play with them, awaken the mule gently first with kind and soothing words uttered in melting tones of voice. This was not done by the perturbed feline. The resulting effect to the cat was the loss of her nine lives at one fell swoop, when propelled by the mule's instantaneous and best efforts she sailed across the street crashing against the wall of a stone building.

"*Pobre gatito!* poor little cat!" murmured Rafael in his sympathetic soft voice as he lifted the limp form by its skinny tail. "Come with me, we will yet find a way by which you can enliven the existence of others although yours has ended."

As I said, Rafael came into Doña Anselma's; there was a slight flip of his hand as he neared the *ollas* of *menudo*. The table was surrounded by a full quota of customers. The hostess with smile and jest proceeded to ladle out the big, deep soup plates full of the succulent life-giving *menudo*. Suddenly a strange and awful change transformed the face of Doña Anselma . . . hanging across the ladle was the furry carcass of one of the most repulsive, squashy, dead cats ever beheld by man.

The screech emitted by the horrified Doña cannot be depicted in print, the language she used would be prohibited in the mails. And when the cat which she hurled at the table of expectant guests landed smack in the face of an able-bodied drunk who, awakened from his serene slumbers by her screams, raised his unsuspecting head to see why he had been so rudely disturbed, the fight started.

When the police reached the scene they were convinced that there had been a fierce cutting affray, for in the dim light the combatants were apparently covered with blood. *Chili colorado* splashed indiscriminately on white suits gives exactly that impression. Old Anselma was eventually soothed, explanations entered into. Rafael, whose white clothes were still immaculate, had absented himself early in the fray, and was quietly ordering a drink at a *cantina* a few doors away. He was heard to remark:

"*La vieja* Anselma is too irascible to carry on a restaurant business as it ought to be conducted. I, for one, intend to transfer my patronage elsewhere."

Rafael, by the way, was one of the most perfect double-jackers in the camp. I doubt if you could find a better one anywhere. When that boy began to swing an eight-pound sledge you got true poetry of motion, rhythm, and timing.

Menudo was only one of many delicious Mexican foods. I have already spoken of the young goat stew made by the wife of the station keeper at Teboreachic. Young goat is every bit as tender and savory as is young lamb, in fact it is hard to detect the difference between the two.

Huevos al Ranchero are a favorite with me. You may use olive oil or *manteca,* hog fat, in which to fry plenty of green peppers and onions, and salted and peppered to taste with plenty of *chili colorado;* when these are cooked to a nicety, put your eggs lightly on the top of the cooked mass and allow them to poach, as it were.

It was great fun on a Saturday or Sunday night to saddle up and ride down town to have some kind old *fondera* make us a nice supper of *sopa de chili verde con queso.* This is a soup, as its name implies; it is made with green *chili* and cheese. We will have served with it a big platter of hot *enchiladas* and a bottle of good claret. It is best to send her word an hour or two before we are coming so that she will not be hurried in preparing the supper, for you are too hungry to wait.

If you have never had the opportunity to eat a plate of good *sopa de queso* you do not know good food. The first time this conjunction of circumstances comes your way do not hold back. The *enchilada* is made from uncooked *tortilla* in which is wrapped a delectable combination of finely ground and highly seasoned meat or chicken with plenty of chopped olives; then it is fried. After it has been cooked this way, the *enchiladas* are laid side by side in a pile three or four inches thick and covered with grated Mexican cheese, which is somewhat like Parmesan but not quite the same, and lastly they are enveloped in a sauce made from finely chopped onions, *chili colorado*, olives, and a few other enticing-tasting tidbits. You put them on your plate two piles at a time or even more if your companion eats fast. With your fork you cut off about a fourth and proceed as customary. When you get this far you need no further instructions from anyone.

I am so fond of *enchiladas* I forgot to tell you how to make the *sopa*. Procure as many green peppers as you think you can eat, cut them around the pepper in pieces about half an inch wide, stew this in enough water to cover it and yet to allow it to have some juice. When it is about done, put in the cheese, not grated but cut into small bits about half an inch square; allow this to simmer until the cheese gets soft. You will find that the dry Mexican cheese will not melt as rapidly as does our yellow American cheese. Take your spoon in your right hand. . . .

If there are no spoons, take your *tortilla*, tear this into four equal parts, and use each quarter as a scoop, eating the scoop when desired or when it becomes too soggy and soft to function as a scoop. The *sopa* being related to the international soup should be eaten as a first course as custom demands. It is perfectly proper to drink some of the claret with this kind of soup, the rest with the *enchiladas*. If you do full justice to these two comestibles there will be no roast.

Frijoles are beans. And *frijoles* prepared and seasoned by a good Mexican cook make a delightful dish. The same thing may be said of the white army bean when prepared by some French women cooks up in the Vosges. It is apparently not so much the bean itself as it is the artisan who manipulates it. Whatever a great artist paints, is a picture that looks like something. The poor artist turns out exactly what the poor cook does . . . "just beans."

While we are on the subject of foods, let me introduce you to the best spinach in the world, which is not spinach at all.

It is a weed which springs from the earth to bless the stomach of man at the start of the rainy season, and, therefore, can be enjoyed for only two or maybe three months during the year. I shall not attempt to instruct you how to cook it, for you never could do it in an aluminum pot or frying pan on an electric stove. It has to be done by a Mexican woman squatting down over a small wood fire, cooked in a baked-clay *olla*, with the knowledge of the right soupçon of onion, lard, and salt to make it what it should be. You need not look over your files of the *Home-makers' Magazine*, no one has ever advertised *quelite* as a vitamin-filled food; I do not believe that *quelite* ever heard of a vitamin. Transportation has been very bad down there and the vitamin, like a lot of other modern things, has probably not reached into the country as far as the Sierras.

The Mexican is interested in food simply and solely as food and the pleasure it gives his mouth and his belly. He is an ignorant cuss and has no particular use for color schemes and such things in food; he would undoubtedly tell you: "What is the use of color schemes in food? Foods are only helpful when consumed, and who the Hell sees what is in your stomach?"

One occasion, when I was a young man, we had been having a very bad time on the trail, as I accompanied the *conducta* into Chihuahua. It was the rainy season, and we had been delayed by roads heavy with mud, swollen streams, bad crossings, and all its concomitant ills; our grub had given out when we reached El Fortin tired, harassed, hungry. We were twelve men on this especial trip with the stages of the *conducta.*

The proprietor at El Fortin greeted us with the news that he had nothing to sell us but dried beef and some *frijoles*, but that his woman would make us some *tortillas*. No chickens, no eggs, no young pigs, nothing but the eternal *carne seca* and *frijoles*. He did have some *tequila* and with that potent drink to sustain us, we got our three teams of fourteen mules under shelter and fed. By that time our reliable old cook Antonio had concocted a stew of dried beef and *frijoles*, which we promptly demolished. The *carne seca* was particularly tough and tasteless, but we paid it scant heed at the time, only thankful enough for anything to fill the empty void.

Early in the morning of the next day, we were refreshed and pulled out from El Fortin. I took with us a poor devil with a sore foot who wanted to get to Chihuahua for medical treatment. I gave him two or three smokes and a drink of *tequila*. He asked me if I did not find the *carne seca* which mine host at El Fortin had sold us rather bad. Knowing that this question had something behind it I agreed that it was worse than usual. I asked him where it had come from.

"*Carne de burro*, meat of a donkey," he answered with a wag of his head. "And not only that," he further volunteered, "but a dead burro."

"Naturally it had to be a dead burro or else how could it be made into *carne seca?*"

"Yes, that is quite true," he agreed. "But it is a sin to sell anyone the meat of a donkey who died of a disease; and these eyes of mine saw what a very sick donkey that was when he died."

It seemed to me to be time for a big drink. Though I rarely ever took one before breakfast even in those robustious years, that morning, on account of the rains and the burro—well, you know how such things are.

There used to be quite a business in Mazatlan in the livers and certain other parts of sharks (seems to me it was the fins) to be sold to the Chinese in San Francisco. The sharks were caught by the natives there on the coast. I had the opportunity to see them at this doubtfully safe occupation one time. Some of the shark fishermen were in dug-out canoes and those canoes appeared to me to be very narrow ones. I was told that only an Indian could possibly navigate them and not turn over.

Although I have been in many places of danger and have seen a great number of other people in like predicament, I never got anything but an unpleasant thrill from it.

When Bob Wagner and I, in a seaworthy boat of ample proportions, watched those Indians hook a seven-foot shark and work him up the edge of that little dug-out canoe, and then club him to death, it was borne in upon me that if it took courage to do this, I was entirely confirmed in the doubt that had always been mine as to whether I possessed any of that particular brand of moral excellence. Each of the canoes was manned by two Indians; as soon as the shark was hooked, one of the men deserted his own line in order to devote his time and energy to working the canoe with his paddle while his partner attended to wearing out the shark. The fishing for these sea beasts was done off the mouth of the harbor at Mazatlan where there was a very respectable swell, and the way the two men would handle that infernally difficult situation was a thing to arouse the greatest possible admiration and respect.

What would have taken place had both men hooked a shark at the same time, I will let you imagine. When I asked that question of our boatman, he shrugged his shoulders with a "*Que sepa Dios*," God knows.

The men had no poles, no reels, no refinements of any kind, and when the shark ran (or whatever you call it when he leaves that especial locality), the men paid out the line with their bare hands. I saw one fellow break it gradually by holding the line down to the bottom of the dug-out with his foot. He wore a *guarache* it is true, but the line was almost smoking hot as it paid out over the gunnel of that old piece of a one-time monarch of the forest. Of course, it was a heavy line. They exhibited wonderful skill, really, in adjusting their weight to the necessities that arose in order to keep the canoe right side up. While they were very earnest in their work and exchanged a shout of some evident import one to the other now and then, there was no apparent haste nor excitement about any of their actions.

I made a trip one time on a little steamer whose forward hatch was filled with dried shark, or fish, or some other sort of sea-faring animal. Most of the voyage was made in very calm weather, and we got the air movement from the bow to the stern with no relief for many days. The old hooker sweated out her eight or nine knots an hour and I wondered what the Chinese would do with the cargo when it finally was delivered to them in San Francisco. 'Way up in the bow was the only spot one could get any air that was not seventy-five per cent contaminated.

Habits and customs in foods are as varied as the thoughts that inspire them. Misinformation as to the wishes of the poor *peon* in Mexico for different food from what he had was rife at one time in the publications in United States. One thing among the mass of erroneous information is vivid in my memory. It was to the effect that the writer had come in contact with two poor women, grinding their *nestamal*, which is something like our lye-hominy (or samp or big hominy—it is all the same). The women were grinding the *nestamal* preparatory to making their *tortillas* from the paste obtained in this way. It is done on a *metate*, a flat stone especially shaped for that purpose with the aid of another smooth stone somewhat like a short rolling-pin. In an earthenware pot the *frijoles* were cooking. To the inquiry as to what they were making, and finding that it would be bread: "What, no flour? No wheat flour with which to make your bread?" "No, ah no! we never see flour, all we have is corn and beans. Meat we have every now and then."

Now, that may sound perfectly dreadful to those who have never seen nor eaten a *tortilla*, some people shudder at the thought of corn bread. If you tried to make Mexicans and

Indians—except a small proportion of the city dwellers—live and work on a wheat-bread diet they would kick like a couple of bay steers. The *tortilla* and the *frijoles* with their *queso*, and a small quantity of meat now and then, keep them in excellent health. Where there is irrigation, and, as I have said, during the rainy season, they have vegetables. This food is what they like (do not the Afghanistans eat sour mash from mare's milk?), what they have been brought up on. Frequently in bad years, our company paid as high or higher than six dollars a bushel for corn and had it imported from the United States in order to keep our people contented, which they would not have been if we had supplied them with wheat flour. They will gladly pay more for corn with which to make their beloved *tortilla* than to make and cook wheat *tortillas*.

It was not only misinformation about food that went to readers of the press in the United States. Once I boarded a train at Chihuahua en route to El Paso, Texas, and ran into a couple of men I knew from the States who were on their way home. They introduced me to a lady also on her way back to the States after having made a tour of Mexico. She was a very plump lady, and excessively wrought up over many annoyances she had encountered in Mexico. One thing that was burning her up was the difference in foods between her Ohio home town and those of Mexico, which food was not prepared properly.

Her chief complaint which gave her no rest was the clothes worn by the poor *peon* in Vera Cruz and the surrounding countryside. The poor peon went about in a pair of cotton pants, or long, loose drawers, a single cotton shirt, a rough straw hat, and a pair of *guaraches*.

"It was simply appalling," she stated. "And oh, the heat! It was so hot I nearly suffocated, I almost died, dreadful, dreadful!"

This last grievance gave me my chance. I asked her if the poor *peons* in their cotton drawers seemed to be suffering from the heat as she had done.

"Oh, no, they didn't seem to mind it at all."

"Then, my dear Madam, when you make your next trip into Vera Cruz why not try wearing a pair of cotton drawers and a cotton shirt? I am sure you would not suffer nearly so much from the extreme heat."

"You are utterly disgusting," was her offended reply.

I made my escape marveling at the attitude of mind shown by many travelers in the criticisms they make of the food, customs, clothes, beliefs of the natives in the countries in which they are touring.

Another thing that struck me as peculiarly insular and provincial, this time in a supposedly cosmopolitan man, was something that occurred in Chihuahua on a certain day when I was going to the States for a vacation. I made a date with two or three friends to meet me at the hotel, after I had changed my traveling clothes, in order to have an aperitif before I attended a formal dinner at the splendid home of Ambassador Creel. This gentleman had been in the Banco Minero when I was unloading the bullion there and had invited me to dine with him that night.

At the hotel I changed the riding clothes in which I had traveled from Batopilas, and went into the *patio* to join my friends. I found them chuckling with mirth. They introduced me to a New York mining engineer who had just arrived on the train from New York, and he was all dressed up in a complete riding rig.

"What was the joke?" I asked Alberto when we had settled down at the table in the jockey club across from the Plaza.

"When you came down the steps into the *patio* in a tail coat and a top hat, this little stranger said, 'I say, look at the tenderfoot, who in the world can he be?' And," continued Alberto, "he is all ready for the ride of fifteen miles behind

Father's best team of English carriage horses; fifteen miles to Santa Eulalia tomorrow when he goes to meet Bill who is selling him his mine. He was all dressed up like you saw him when he got off the train this morning, only then he had a big six-shooter strapped on him and was carrying a Winchester .44."

"Where are they now?" but he did not hear me.

"Here is a ticket to the theatre. Meet us there after the dinner. We will have some fun, and make you miss your train tomorrow. There are dancers from Spain tonight, and they ought to be worth seeing."

I tell this seemingly trivial incident to show how strangely ignorant some persons are about other people and other places. This man was a mining engineer, supposedly having had a wide experience, he was over thirty years old, from New York City, had come all that distance to purchase a mine at Santa Eulalia, which was a carriage ride from Chihuahua. Chihuahua, a city of thirty-five thousand population, was the junction of two railroads. It was more than two hundred years old, through which and around which the Lord only knows how many millions of dollars had come and been produced and had been paid to stockholders of two continents. Yet this person thought that because another man dressed appropriately for a formal dinner he must be a tenderfoot.

Why, Hell! Some people even in those days dressed for dinner in Chicago! If he had been at our camp on an anniversary of some sort, I suppose he would have yelled bloody murder and thought that all of us were insane when he found us dressed in tail coats. If he had had a Scotch and soda that came direct from Scotland in special fifteen-gallon kegs, he would have thought that he was dreaming!

XXIV

El Patron Grande with Angels and Calamity

The *Patron Grande* was the kindest and the most charitable man I have ever encountered in more than sixty years wrestling with what we call life. He had an ample field in which to practice these characteristics in the mining camp. In the first fifteen or twenty years of our existence down in the Sierras occasionally old tramps of non-native parentage would wander in. Sometimes it would be perplexing to the last degree to discover their nationality until they had forced upon them a bath, a haircut, and a clean outfit of clothing from the skin out, and then we would be able to decide with which geological age and continent they had been connected.

Very early in the game these wanderers were called "angels." Dad always said that it was necessary to take them in, "because who could tell whether we might not be entertaining one of the winged divines unaware?" This was confusing for a long time to my kid brother and myself when we were small boys, for we could not tie together the idea of angels as we had been taught to conceive them, and these battered dirty old wrecks.

One of them we called "Roses" and he was never known by any other name. Over a period of years he made us many visits. He would come into camp worn with fatigue, thin, dirty, hungry. He would leave whenever he pleased after a few months, clean, well-dressed, rested, fattened. He acquired the name of Roses one night when he reached the front gate to the Hacienda San Miguel very drunk and singing lustily. He loved, when he had acquired this stage of exaltation, to quote Shakespeare and other authors, correctly or otherwise. The night watchman and my uncle Sam Young (who was a sort of general superintendent of the whole hacienda at night) escorted Roses to the rooms over across the corral where he slept, endeavoring to pacify him in order to prevent his disturbing those of us whose windows overlooked this lower level.

Thanking their lucky stars that he had at last succumbed to their efforts they laid him on his bed and left him to his alcoholic slumbers . . . as they fondly hoped. But no! Out of his room Roses flung himself screaming: "Sit down, Tommy. Sit down I tell you! Listen to me while I tell you that this world ain't no bed o' roses!"

Again my Uncle Sam, who was a gentle man, ably assisted by the night watchman, strove by kind words and calming tones to quiet him. But Roses was adamant in insisting in a whiskey voice again and yet again that this life was no bed of roses.

"Sam," this voice was that of my father, it was an impressive sound. It trembled a little and this denoted high steam pressure, a tone of voice thoroughly understood by all. "Sam, knock that old cockroach on the head and drop him over the river wall."

Absolute silence prevailed. Presently Roses turned toward his sleeping quarters whispering hoarsely: "What did I tell you, Tommy? This world ain't no bed o' roses. No one around here appreciates Shakespeare. They ain't even up to the lesser poets."

When Roses paid his first visit to the camp he must have been around fifty years old, he was over sixty when he passed from our ken to, I hope, that bed of roses he constantly talked about when he had finished his bottle of *mescal*. He was a wizened-up old man. I have not a doubt that to his diminutive size could be attributed the fact that he had not long ago been exterminated by someone. He was a fearful pest. The event I have just narrated occurred in the early hours of a certain Monday morning, and when my uncle reported to *El Patron Grande* before turning in for his sleep, my father said to him:

"Sam, why in Hell do you take so much trouble with that old reprobate? Why don't you knock him on the head?"

"Because, brother-in-law, every Sunday night you WILL read the Church service to all of us and insist on my singing beautiful Christian hymns. [Uncle Sam had a fine voice.] Last night you laid especial stress on the chapter which has to do with charity. What you read in the way of prayers and chapters from the Holy Book evidently penetrates deeper with some of your listeners than they do with the lay-reader himself."

The arrival of one of Dad's angels was as much of a calamity as the possession of a horse by that name almost turned out to be. My father was a large man, six feet two in height; when he reached his early fifties he had accumulated some extra pounds; not fat, but a big man. He weighed about two hundred and thirty pounds. It was not a simple matter to find a riding animal that was up to his weight and other requirements. Uncle Jack helped me one day in making a trade of my ill-mannered pony for Calamity. We paid something to boot, for Calamity was a strong good-sized horse, well-gaited. Very soon after he came into my possession I went back to school at Sewanee, and father began using him and liked him; he was riding Calamity when he came near to having a fatal accident. This near tragedy broke us from giving any name to our saddle animals which might have a sinister meaning.

One of the many pieces of work that was put through at the camp by my indomitable father was the more than three-mile aqueduct of solid masonry built along the side of the canyon. For many feet the leveling was done through hard rock, in some cases short tunnels from thirty to a hundred feet had to be driven. Regular inspection of the work was made; when it was possible to do so the party on inspection and pleasure bent would ride through the aqueduct itself, since its floor made a very acceptable roadbed. On one afternoon my father was mounted on old Calamity and had ridden about half way through the longest of the tunnels where the light was dim. A drunken Indian arose suddenly from the floor where he had been asleep, directly under Calamity's nose. The horse, of course, reared in fright, and threw my father's head against the jagged rock of the roof, giving him a nasty scalp wound which nearly caused his death after erysipelas set in.

I can never forget the feeling of accessory before the fact that was mine when at school in the States I received word of this accident. Was I not the owner of the animal? Had I not acquiesced in my uncle's bestowing the name of Calamity upon him? As a matter of fact Calamity looked it, though that was more on account of his Roman nose than of any meanness of disposition.

Quite often while the work of Porfirio Diaz Tunnel was under way and the heavy masonry work of the Hacienda San Antonio was under construction, the *Patron Grande* accompanied by his son-in-law Walter Brodie would walk the half mile to the tunnel on a tour of inspection. Walter was responsible for that engineering feat. Neither of these gentlemen was inclined to tote a big gun; they usually slipped a small revolver, generally a .38 into their hip pockets. This habit caused Mr. L. H. Stevens, the vice-president of the Company, to send them each as a present a hammerless Smith & Wesson safety. It had a plate on the stock that you were supposed to squeeze as you pulled the trigger, else it would not fire.

One afternoon they were walking to the tunnel on such an inspection when that great piece of work was nearing completion after heart-breaking months of hard times and tribulation. They were about half way there, when a vicious-looking dog darted out at them from a small *jacal* on the mountainside. At the same instant each man drew his little gun and aiming, pulled the triggers and pulled the triggers till they could pull no more, but never the sign of an explosion. Along came a small boy who shied a rock at the vicious beast. This worthy defender of the home fled to the protection of his refuge, the little tyke grinned at the *Patron Grande* and his son-in-law, and went merrily on his way.

I believe the laughs they got from the oft-repeated story was worth hundreds of dollars in doctors' bills and medicines. Both guns were turned over to Marshall Willis who gave them a good soaking in oil and a thorough cleaning. The gentlemen went to the upper *patio* privately afterwards and fired the guns a time or two for their own inward satisfaction.

My father inspired a strong feeling of affection and one of deep-seated respect in his workmen. The only drawback to a complete one hundred per cent sympathy was his Spanish. It was a bit out of the ordinary and difficult to understand. Thus it became a habit whenever possible for one of us who knew what he would be trying to say to be present or within hearing of what was said, so that the *mozo* to whom instructions had been given could slip to the *patroncito* unknown to the *Patron Grande*, and ask him to translate the order. Many times I would come to the main office and see a poor *arriero* standing there scratching his head with a lost-soul expression wondering what in the world he was to do. Then I would go into the office and tell my father that so-and-so had not understood him exactly and would he repeat to me the instructions and I would translate them to him.

"What?" my Dad would say. "Another damned fool who don't understand his own language? Well, I don't blame him, I don't either."

We all caught our bit of Hell now and then. It was usually justified. I recall Clemo particularly in this connection. He was an old Cornish miner and he was a corker when it came to putting in timbers. The only trouble with him was that he did from time to time get beyond his normal condition of alcoholic stimulation and then his work was not so good.

On the other hand, if he became entirely sober he was no good at all. So long as he kept to, say, a quart of *mescal* about every ten hours, he was in fine shape. He taught many of us more than a little about timbering.

Every Sunday morning at ten o'clock the mine superintendents and sometimes their assistants would come to the main office to have individual chats with my father. After this was over, we would all assemble in the big main room around the large table where old Domingo had already deposited a tray covered with sandwiches, cakes, and things . . . not forgetting the Scotch whiskey and the sparklet siphon. For twenty years, every Sunday the *Patron Grande* would turn to old Clemo and say:

"Come on, old man, have a Scotch and soda with us."

"Oh, no, Governor, thanks. I don't believe I will have one today."

"Oh, come on and try it!"

"Well, sir, if you insist I will take a small one."

He would seize the tumbler in his huge fist, taking the bottle in the other, then would engage all of us in deep conversation; before the tumbler got so full that one could see the whiskey above the edge of his hand, he would put the bottle back on the table and shoot what was apparently a quantity of water into the Scotch from the sparklet siphon. With a graceful "Your health, Governor!" he would absorb all but about half the contents of the tumbler in one gulp. Then he would say: "If you do not mind, I would like a little water. I never could stand liquor without water."

There are fifty-two Sundays in a year, sometimes more; I am willing to gamble that old Don Guillermo Clemo never failed once in the ritual and I actually believe that he thought he fooled all of us each and every time. Once or twice the *Patron* would glance at us and apparently forget to urge old Clemo when his turn came. After a perceptible pause, the old fellow would say: "Well, Governor, if you insist, I believe I will," and he would put on the same show not one whit abashed. Just the same, the old Cornishman was all there. Many times working with him in some damned hard piece of timbering job I would thank my stars that I need have no fear that anything wrong would escape him.

Because in all companies there is a director who is always a perpetual suggester, and faultfinder, our company was no exception. My father had a letter from ours telling him that he believed there was a superabundance of sons, sons-in-law, nephews and brothers-in-law on the payroll. This director received a reply by telegraph from my father to the effect that if he could supply others who would do the same amount of work on the same salary, please to send them down to Mexico. He wanted it to be very plain that he, the *Patron Grande*, was ashamed to employ trained men to whom he was obliged to pay so small a salary for the quantities of work turned out, but that the company could not pay larger salaries unless the stockholders would agree to an annual assessment. The suggester was formerly a native of Scotland, and had more than once been cautioned for disturbing the peace by squeezing the eagle on the dollar until he screamed; therefore, we heard nothing further from him on the subject.

Most of the directors of the company played the game. But it was only natural that because at the beginning a couple of very big bonanzas were taken out, the uninitiated among the stockholders could not understand why this heavy production could not be a constant one. They could see no logic in the explanation that these rich bodies of ore lay distances apart and were hard to find, that there was hard rock to be broken, and many other difficulties to be overcome before another one could be reached. These first bonanzas were the reason for many headaches; they caused a prodigious error in calculating what might be regularly expected in the future. Therefore, the improvements were started, the cost ran into the millions, no sufficient capital subscribed to cover them. Hence the mines had to take over this burden from their regular monthly production. This resulted in a strong desire to take advantage of any big production era to sell the property.

We had no objection, for we felt that we could use our share to as good a purpose and could not have to work nearly so hard in some other part of the world. All this resulted in visits upon three separate occasions by representatives of syndicates who came to investigate and examine the mines with the object of buying the property. The Spanish War killed one opportunity to sell; the Boer War killed another; and the Madero Revolution settled the affairs of Batopilas.

A party of these experts direct from London spent a Christmas with us. An English engineer came with them whose reputation and experience embraced South Africa, Canada, and God knows where! He arrived in Batopilas as second in command, the real underground man, one of those supermen, as it were. He could look into and through hard rock and tell you all sorts of things which you found to be wrong after you had laboriously broken the ground and weeks later had come to the place where these things, according to his X-ray eyes and supernatural intuition, were supposed to be.

Por mis grandes pecados it was my misfortune to be in charge of the particular mine upon which he elected to make his first examination. After a preliminary talk with him I was duly impressed with his vast experience and superior knowledge. "For," said I to myself, "this man would not talk so high, wide and handsome unless he was as good as he says he is. Otherwise he would be shown up *pronto*."

Commencing on Monday I was to take him over all the old workings of the old San Miguel as well as those below tunnel level. This job, if it was to be a real one, would entail a couple of weeks. The old workings were a thing of real magnitude, especially since all the ladder-roads except one main set had long ago been taken out to be used in the present development.

I explained this to him at great length. I asked him to give me a week to arrange a semblance of ladder-roads, else we would be compelled to do a great deal of climbing up and down ropes. Most of the veins were of a dip that permitted climbing hand over hand while partially walking on the foot wall, but it would be a tiresome business. Unless one was used to such things it would be dangerous because we would have to cross stopes which were several hundred feet high and where a misstep would result disastrously.

The English engineer ridiculed me and said: "Of course, if you are afraid to do it as it now stands, why, I will wait till you prepare the ladder-ways, but certainly there is no need to do so on my account." Then he recounted to me some of his extraordinary exploits in the mining field. So I shut up and let it go at that.

We were all ready for him on the appointed morning. I had engaged to accompany us two of my best climbers who knew every working in the mine, and we went in.

I had tried to prevail upon him to put on one of the cheap straw hats we were using, because one can safely stick the hook of the miners' lamps then in use through the bent-up hat brim. This he refused to do, saying that he never needed to use but one hand when he was climbing and, therefore, could carry his lamp in the other. I had looked askance at his stiff new boots, which were not only uncommonly heavy in the soles, but were laced up tightly. Since I had told him before what I considered the best thing for him to wear on his feet for the work in hand, and had been severely snubbed for my pains, I refrained from further suggestions and we entered just as he was.

We went up the old ladder-road to the five hundred foot level, called by the name of some saint or other, the invariable custom of the Mexican miner. Before reaching this level I knew by his awkwardness that we were in for it, but I hoped that he would learn something about how best to do it by watching the men who were ahead of him. Wherever it was possible, these men from then on climbed and swung about to put ropes in places for hand-hold. With laborious exertion we crept along and had reached a point by noon where we would be able to get to tunnel level by going down a hundred feet, and there we could have our dinner. Long ago the visiting engineer had lost his lamp and was in a fearful sweat nearing collapse from fright and exhaustion. Thankfully he agreed to my suggestion about getting to tunnel level to have our dinner.

I said: "Will you go ahead or shall I?" He gasped out: "Let me go ahead!" He did so, his feet slipped, he hung suspended by his hands. This was a matter easily remedied, all he had to do was once more to place his feet on the foot wall, let his buttocks hang down with his legs at right angles to his body, and thus gradually work them down, while he let go the rope in a hand-over-hand manner. But he was through and could only pant and hang there. I had to climb over him, get my shoulders under him, and by degrees we worked ourselves the hundred feet down to tunnel level.

We had dinner outside.

After a rest for digestion, I said: "We can go back now and finish our investigation."

He swelled all up and told me that for the present he had seen all that was necessary. So help me Bob, that is the last we saw of him at San Miguel Mine except on tunnel level, where he spent two or three days fooling about to establish an alibi.

The same thing happened at the other mines. We had a perfect set of maps which were absolutely up to date. He had access to the records of the Assay Office for the ore values of mill-runs for twenty years and the bullion records. He really had a much more comprehensive and completely thorough record than he could have made himself in a million years. His report was so very flowery when it was submitted by the chief of the expedition to my father as a work of art, that the *Patron Grande* after he read it refused to send it to anyone.

"The report is entirely misleading. You have theorized as to ore values and credited possible returns that twenty years of actual work showed were impossible. The lower grade ores would never average what your report claims for them."

"Well, but did you or did you not want to sell the property?" blustered the chief of the expedition. "Did you or did you not arrange to have us make this examination?"

"I did not," stated my father. "I knew absolutely nothing about any of it until I heard from New York that they had overtures from an English company saying they might be interested in buying the property. The Board of Directors agreed that they would send engineers down here. My figures on the books are a true record of twenty years, and they are good enough to sell the property on. I will readily agree for you to copy these figures since they are without embellishment. But I will be eternally damned if I will endorse even by silence such fool statements as those that have been made by your engineer and O.K.ed by yourself. I do not think that any of you know enough about mines and mining to warrant your having a position in my mines even as an assistant to any single one of my superintendents or mine bosses."

As I have intimated previously, my father was a forceful man.

He had instituted a company store among all the other things. If this had not been done the native and the *gachupin* storekeeper would have raised the prices of all commodities to such a height that it would have been impossible to give a man a living wage, and at the same time to keep the mines in operation with the price of silver where it was.

Wise President Diaz gave permission to have the company

store established at the camp, being satisfied in his own mind that my father was interested in the welfare, physical, mental, social, moral, of our workmen. Our plan was to stock the store with everything from bread and beef to needles and shoes. When a man went to work on Monday morning his family could draw an order on the store for the sum that his half-week's earnings would be. This insured sufficient food and other necessaries for the week. Thus we did not care nor did his family suffer if the miner blew the balance of his wages received on Saturday night on *monte* or *mescal.*

It was an innovation; there was some grumbling in the beginning but being of a philosophical race they soon appreciated the advantages of a well-fed family. The results obtained in general health and improvement in other ways were truly remarkable and gratifying. The company was in a position to buy in large quantities and thus to obtain lower prices; we sold as close to actual cost as it could be done; the proposition worked out successfully, even though it was a great nuisance to my father and much extra trouble. The company considered it best to return to the old system after trying out the store for a few years, but after making a general inquiry among the men and their families as to how they felt about it, such a howl went up against any change, and requests to keep to the present method to save them from their town merchants were so urgent, that we decided to keep on with the store.

Often under the old way when a workman was paid off on Saturday he would reach home with little or none of his pay left. This would mean a hungry, therefore discontented, family; it would mean that the man showed up to go to work on Monday morning with an empty belly and no lunch. Now, these two sources of irritation are detrimental to a good day's work, there is a strong tendency towards becoming quarrelsome . . . and that means trouble. Or it would mean his getting in the clutches of some native or Spanish storekeeper who would cheat him to the Nth degree. The storekeeper would be a tiresome problem on pay day to the company hanging about the pay-window waiting for the workmen to be paid off. Under our store system, there were no empty bellies, there was greater contentment, health, cleanliness, and a well-clothed community. I quote from the policy of my father's friend, the clear-seeing ruler, His Excellency Don Porfirio Diaz. . . . It is much more difficult to interest a contented family man in any change such as a revolution may bring to them, than to interest a discontented man. The greatest reasons for discontent are an empty belly and idleness. Work and a full belly make things run along smoothly.

XXV

Death of *El Patron Grande*

In the late summer of 1902 my father died. At the time I was at the country place Bleak House in Washington still confined to my bed after a severe operation.

Notification came to us of the serious illness of my father, and my mother immediately set out on that long and trying journey to Batopilas escorted by my cousin, Dr. Merchant.

She made a very rapid trip; everything was done by our good friends in and out of Mexico, officials and others, to speed her. But my father had passed on before she could reach him.

As soon as the surgeon permitted me to travel, I returned to Batopilas; this was in October, I think.

My father's body had been placed in a tomb on the mountainside above the *hacienda* where he had established a small cemetery: there also had been entombed two infant grandchildren.

It was my mother's desire and that of the other members of the family that his body should be finally laid to rest in Washington, the place of his birth and the scene to which he had devoted the earlier portion of his life. Washington was the place where he had unselfishly built up, against unscrupulous and violent opposition, the nation's capital, preventing its removal to some interior section of the country, by which move the traditions and desires of the patriots who had founded the nation would have been discarded in order to satiate the greed of those politicians who came on later, after all the hard work had been done—those politicians who wished to take to themselves the benefits which would undoubtedly accrue to the town and who might have been able to grab off the plum by use of crooked political schemes.

A heavy casket is a difficult object to be transported across one hundred and eighty-five miles of the Sierra Madre Mountain trails. There was but one way that this could be accomplished and that was on the shoulders of strong and devoted men.

And so it was done.

In the doing was manifested the true respect and affection this community of Mexican citizens held for the man who dedicated all of his magnificent abilities and all of his time for twenty-two years to building up a business which employed as many as fifteen hundred men; who paid them a minimum wage of one dollar and a half a day for unskilled labor; who protected them by preventing unfair advances in the cost of the necessities of life; who gave them and their families free hospitalization and medical attention; who founded and established a happy and contented community of approximately six thousand souls.

We came to the conclusion that it would be necessary to have the carrying cradle for the casket so arranged that two men could support it at the same time at each end, and two on either side . . . thus making eight carriers. The Mexicans are accustomed to conveying loads on their backs and shoulders; they are very remarkable in this way. They fold their *serape* in such a manner that when they are hung on the *necapal*, which is a short broad leather strap with the ends connected by leather thongs, the strap placed on the forehead, the *serape* will form a pad on the shoulders at about the level of the top of an Infantryman's combat pack. The handles of the carrying cradle rested on these pads formed by the *serape*, and the carriers moved forward at a steady walk.

We had five sets of eight men who worked as a unit, each set matched as nearly as possible as to size; the groups relieved one another at intervals of twenty minutes to half an hour while in motion. Thus there was no loss of time on the trail and the casket was placed on the ground but once during the day's march of thirty-five miles, and this was when the noon rest was taken and food eaten.

We made the one hundred and eighty-five miles on the regular *conducta* pack-train schedule . . . that is, in five days.

It is not a task that would appeal to most men even for good pay; to be compelled to march thirty-five miles each day over rough and narrow mountain trails for a distance such as that trip consumed. Yet when it became known in Batopilas among our workmen that the journey was contemplated, many, many times the number of men needed, volunteered to carry the casket. They were careful to specify:

"Without pay for the trip. Solely for the desire to demonstrate the respect and affection for *El Patron Grande* and *Doña Maria* his wife, who had in all ways done so very much for us and for our children."

Had it not been for the many years of experience and the smooth-running organization developed at Batopilas, to make such a journey would have been next to impossible. As it was, we managed it on regular fast-mounted travel schedule and with little or no trouble.

The pack-train with the beds and the food set out ahead of us the first day, with sufficient start to enable them to be already installed at the end of the day's march at the Station, with the supper well on its way towards readiness for those of us who followed with the casket. More than forty men, of course, require a great deal of food and good substantial food. Fresh beef had been arranged for by the station keepers who had been advised by my brother Conness that this must be on hand. Old Antonio Navarro, one of the most faithful and sure, was in charge of the kitchen.

On the second morning of the journey, my brother Alex and I and the carriers with the casket started shortly after day break; before we reached the point where we took our midday rest of an hour, the pack-train overtook us and passed. At the rate of travel they could maintain, they were at the station at least an hour before we arrived. And thus for five days the cortège moved forward until the last morning. We set forth on that morning long before daylight, because we had not only to reach Carichic early in the afternoon but at that place we transferred to the wheeled transport, and had to drive an additional sixty-five miles. It is safe to say that by trail and by wagon we covered ninety-five miles on that last day.

While we were crossing the mountains, in order that those who were one day in advance of us might know that all was well and that we were keeping up with the schedule, once or twice after reaching our night stopping place we secured a runner from the station keeper. Maybe this messenger would be one of his own sons on horseback. This one would leave immediately upon our arrival at the station to travel fast to the next one ahead where he would arrive some time before midnight. He would have a note from Alex containing the news of our progress and with a reassuring message for my mother. This note was delivered to Conness as soon as the messenger reached the station, and he would give it to my mother in the early morning when that brave lady awakened.

During this her last trip across the Sierra Madre Mountains, in the same manner as on many of her former ones, my mother was carried the greater part of each day's journey in a chair. This was made feasible by arranging a wicker rocking-chair with handles such as were employed in the sedan chair of olden days. Eight good men trained to such work bore the burden two at a time with frequent changes. And this more than tiresome trip was in this way somewhat alleviated for the ladies.

Comparatively few men have ever trudged thirty-five miles a day for five consecutive days in succession, and the feat performed by those fine faithful souls who carried my father's remains across the Sierra mountain trail can be but little appreciated. Always cheerful, never a complaint, never an attempt to shirk when their turn came to assume the burden . . . it was one example from out the vast number of others which I had been a witness to since early childhood that endeared the Mexicans, especially the Batopilense, to all of our name and race.

I suppose it is natural that at times there comes a homesickness for the land and for those people which is almost too strong to be resisted. A yearning which forces one, for the time at least, to regret that circumstances prevent the return for one's few remaining years to the place where was passed early youth as well as a great part of one's manhood.

My mother and other members of the family left under the escort of my brother Conness one day ahead of us; arriving at Carichic they rested for a day at the home of our good friend Don Ignacio Ortega. Alex and I with the casket were met by the largest wagon just beyond El Alamo ranch.

We drove from there into San Antonio, the railroad station, that same day, where a private car was waiting for us. By the time my mother and the others of the family reached there, the casket was settled in its position in the observation end of the car where it rested until we arrived in Washington.

There was an interlude at Chihuahua while we were being transferred and attached to the regular north-bound Mexican Central train. There we received visits of condolence from representatives of the families of my father's devoted friends—Don Luis Terrazas, Don Enrique Creel and his charming wife, Don Antonio Prieto, Don Juan Creel, and many, many others, not only Mexican but other foreign friends. Quantities of flowers of every kind imaginable were brought to the train, and put in the private car.

The trip to Washington was uneventful, and on arrival there an impressive cortège escorted my father's remains to Rock Creek Church, where they temporarily rested in the tomb of a friend, until my father's mausoleum was constructed, in which they now lie with those of my sainted mother.

XXVI

Batopilas without *El Patron Grande*

It was not all beer and skittles during the thirty-odd years of mining the precious metal in Batopilas. Silver went 'way down in price; there was a heavy floating debt and the overdraft allowed by the Banco Minero in Chihuahua was overreached. But before matters hit the absolutely fatal spot at any one time, we would run into a bonanza of greater or less extent, and sometimes in the very last two weeks of the month, at the end of which the bullion would leave for Chihuahua, we would knock out anywhere from fifty to a hundred thousand dollars. One month we took out four hundred and fifty thousand dollars worth of silver. At various times we had run into debt with the bank for around eight hundred thousand dollars and paid it back all from a few months' production.

After my father died, we had been going through a trying financial time. Our credit was not as good as his had been; we were the younger generation and had not proved that we could move mountains as he had done.

My much-beloved brother-in-law Ned Quintard had died suddenly in Washington while we were on the road with my father's remains. Alex had been made general manager at the mines. On a certain afternoon instructions were sent me to report to him at the Main Office. When I got there I found Conness also at the Hacienda San Miguel, and we were informed that the Banco Minero now refused to permit us the overdraft of sixty thousand dollars which had been agreed on, unless we would send them that much as well as a pretty large additional amount in bullion at the end of the month.

Dios es grande! It chanced that the previous night I had showed up in a small working some signs of native silver. This ground *ought* to prove productive. I personally was satisfied in my mind that it would do so. We decided that we could telegraph the bank that we would meet their demands.

We met them.

By the end of the month we sent out a little over one hundred thousand dollars worth of silver bullion. This was the beginning of the bonanza which paid off all the floating debt and more besides during the next twelve months. If I am not mistaken, and I believe that I am not, the debt was eight hundred thousand dollars or thereabouts. This example is cited to demonstrate the richness of the ores of Batopilas and also that we worked plenty hard to get them out.

We had no air drills then. On the tenth of the month we began to take out one hundred thousand dollars in twenty days. We were in a small narrow working, badly ventilated; the ore showed up right down at the foot of the face.

Did we work! We drilled and shot round after round of dynamite in powder smoke so thick that in order to see the head of the drills we had to hold our lamps within a foot of them, and even closer. God! the headaches! No headache is more intense than the one produced by dynamite fumes. Days, nights, days, nights, no sleep except what we could snatch back in the drift every now and then for maybe an hour lying alongside the tracks. . . .

Alex worked far too long and far too hard without a vacation. Finally we prevailed upon him to take this long-deserved change. I took over the management during his absence. He was unavoidably detained two or three times on his journey back to Batopilas from the East, and the last delay was caused by high rivers in the mountains from unusually heavy rains. He had his family with him and could take no risks. Of course, I could not leave the mines before his return. There were some appointments in New York City, of great personal importance to me, which I had been forced to postpone several times by telegraph. The persons with whom the appointments had been made became insistent and almost unpleasant about my procrastination in getting there. New York residents who abide in that city in comfort and financial ease, and who can arrange matters for months ahead, sometimes are a trifle hard to reason with.

At last Alex reached the camp after great inconvenience to himself, in order to make it possible for me to turn things over to him and get away in time, by dint of utmost speed, for the appointments in far-away New York.

He arrived at Batopilas almost exhausted in the early morning after having ridden nearly all night. I turned things over to him (there were innumerable details to clear up), and got away from Batopilas at two o'clock the next morning. I left the camp in a downpour of rain, knowing that the Arroyo de las Huertas, three miles up the river and directly above the dam, had risen very high. I must be obliged to swim the Arroyo, which was two hundred yards wide at that point.

I took some lead-animals with me so they could spell each other, two pack-mules and one saddle-mule. These I sent ahead of me with my loaded pack-mules the afternoon before. Alex and I had a farewell stirrup cup at two in the morning, I bade good-bye to him and some of the boys, and accompanied by my old friend Angel Gil who insisted on going with me as far as the Arroyo de las Huertas, I set forth already weary and anticipating tiresome traveling.

I was riding a powerful sorrel *macho*. A *macho* is a male mule. This animal was a *burrero*, the opposite cross to the usual mule-producing cross. He possessed a good bit of his donkey mother's stubborness with an endurance and strength that was almost supernatural. His sire was a stallion of excellent blood; this *macho* was as strong as a steam engine, courageous, and could swim like a Newfoundland dog. We got to the Arroyo de las Huertas and found it pretty high. I said good-bye to Angel, rode up the bank as far as I could and started into the rolling, swirling current. The minute we got into the water where the depth made the *macho* swim, I slipped from his back, caught hold of his tail and we made a safe landing without being swept down into the Batopilas River. This would have been, to express it mildly, inconvenient, because we then would have been in the body of the water formed by the dam and it was still pitch black dark.

The *macho* and I shook ourselves, I had a pull at my bottle, and we went on to get dried out by the sun, for the day cleared as we advanced.

We climbed the hill out of the Canyon onto the trail to Teboreachic Station, which you will recall is the last night-stopping place on the trip coming over the trail towards Batopilas. Here the *macho* and I joined the pack-mules and the *mozos* with the spare animals. We were all of us outside a big meal very quickly, I with plenty of hot strong coffee, and had two hours' rest. Off again then, mounted on a fresh animal en route to La Laja seven hours' ride away. At La Laja there were four hours' rest and sleep. Seven more hours of riding brought me to Pilares, but before reaching that place I had had to swim the Eurique River. At Pilares I had two hours of rest and moved forward once more.

In good time I reached Gauhochic and found the river fordable. "*Dios es grande*," I said, because when that river is high you absolutely cannot fool with it—exactly below the ford the stream is enclosed by precipitous rock walls with no other bank for many miles. We fed there and moved on . . . at any moment the river might come down upon us, there was no time for rest. If the river rose it would lose me forty miles and the trip must be made on schedule. The stage would be waiting for me in the early morning, and if there were no delay whatsoever I would be able to make the train at two P.M. at San Antonio Railway station and thence to Chihuahua in time to catch the Mexican Central out that same evening en route to El Paso.

I did a hundred and eighty-five miles in the saddle in fifty-two hours. With no margin to spare I made my connections. After I boarded the Southern Pacific Railroad train in El Paso, I had the first opportunity to shave, bathe and change to a costume more appropriate for travel on the Sunset Limited . . . a truly de luxe means of transportation, especially when contrasted with the last fifty-two hours.

I had to walk through two Pullmans in order to reach the drawing-room which I had reserved by telegraph. I was certainly a tough-looking customer, and I received some startled and disapproving glances on my way through the Pullmans . . . the hotel porter lugging my two big bags, I following unshaven, hollow-eyed, still wearing the clothes in which I had left camp, my six-shooter strapped around my waist.

There were some ladies from the East on the train with their husbands on their way home from a trip to California. The earthquake and fire at San Francisco had just devastated that city. Among the passengers was Miss Kathryn Gray the actress, who was returning post haste to the East to fill an engagement. She had made a dash across the continent to San Francisco at the first news of the catastrophe, because her mother was there at that time. It was introduced to Miss Gray and at dinner that night she informed me that I had been the cause of no little speculation on the train after it had drawn out from El Paso. She heard that a disreputable-looking person had got on the train there, and that he must be secreted somewhere on board, since he had not been seen afterwards; he had gone into the drawing-room from which I had emerged later.

It was not until the conductor had told one or two of the passengers that I was the disreputable-looking specimen, that often in this part of the country a haircut, a shave, a bath and a change of clothes produced the same result had they been relieved of apprehension.

I was far from well; had been suffering for a long time from malaria. I saw a doctor in New York who had been highly recommended to me. Among many other ailments he found that my heart was affected; that I was on the way to a dangerous condition of anemia; that I should dose myself with the most immense quantities of various kinds of medicines, go on a strict diet, refrain from anything alcoholic to drink except one pint of Burgundy at lunch and another after dinner. I paid the doctor, I even thanked him. I drank the Burgundy, I like good Burgundy. To that I added many other forms of liquid refreshment, and ate everything delicious that I could buy at Martin's, Delmonico's, many of the hotels, and all the good after-dark places. In sixty days I was slightly over normal in weight and I could push over a house. . . .

XXVII

General Porfirio Diaz

Some years ago, it was popular to declare Porfirio Diaz possessed of all the worst attributes possible for the Chief Executives of a nation to be afflicted with. But if you have lived in a country from the age of five years, from 1880 to 1911, during an era—according to the press of those days and some of its politicians—when it was a hotbed of disorder, peonage, corruption, and a thousand other civic ills; when you yourself have traveled the country, have known hundreds of its people, spoken the language as fluently as you do that of your native country; when you yourself, I say, have never in your life come into contact with a single one of those frightful abuses attributed to Mexico and the Mexicans at that time, but have seen a nation advancing in every way as rapidly as did our own Southwest, you are filled with amazement and a boiling sense of injustice.

Without hesitation, I say that General Porfirio Diaz gave the Mexican nation an era of order and of good government far above anything that we can approach now. There were fifteen million souls in Mexico. There were about fifty different tribes of Indians, each having its own dialect; among many of them only a decided minority could speak Spanish. Inaccessibility was the principal feature of the country.

Not only our company but many others made regular monthly bullion shipments, which ran in value from one hundred to two hundred thousand dollars. These same pack-trains which carried the bullion brought back to the mining camps from fifty to a hundred thousand dollars *in currency* for payrolls. I have already described the journey and the terrain over which the trips were made. In thirty years only once was an attempt made to hold up one of them—that was the monthly *conducta* from the mines of the Pinos Altos Mining Company—and it was unsuccessful.

It was not until the despotism of President Porfirio Diaz was overthrown that our company ever lost a bar of silver or a dollar of payroll money. The great liberator, that oppressed Francisco (Pancho) Villa, stole thirty-eight thousand dollars worth of silver bars and buried them in the *patio* of a house in Chihuahua. He was afraid to convert them into cash, and later when this leader was not so strong in power we recovered the treasure.

All this talk about peonage! How many people knew that the word "peon" in Mexico means an unskilled laborer, a pick-and-shovel man.

It angered us more than once to hear verbal statements from, and to read accounts by, persons who believed that General Diaz made no efforts toward establishing schools or inaugurating sanitary and health reforms.

In the year 1885 I attended a School Commencement at that small mountain camp of Batopilas three hundred miles from a railroad in the very heart of the Sierras, and it was as well conducted as any that could be found anywhere at the same period.

More than forty years past the smallpox epidemics among the Indians and the town inhabitants were nearly eradicated in this way: every medical man anywhere and everywhere in the country was supplied with vaccine and ordered by the Government to vaccinate any and every citizen. The Indians feared any such innovation with a fervent and superstitious terror; the doctors were given a detail of police and soldiers to capture the prospective victims and to hold them while the vaccination took place. When they came to town, they were grabbed and held fast in abject dread and mental torment while the doctor performed.

Our surgeon at this time, Dr. Merchant, rode a big white

horse. I remember one Saturday afternoon watching him galloping down the street, El Calle Principal, with soldiers in hot pursuit of ten Tarahumare Indians on foot, and as many more mounted on their small ponies and *burros,* fleeing in desperate fear of their lives. They made a frantic effort to escape while the unsympathetic populace shouted in delight, encouraging surgeon and police and Indians indiscriminately. The poor wretches were captured at the edge of town; someone headed them off. There they were surrounded and for the good of their souls and bodies they were firmly held and the surgeon cleaned and scraped and inoculated.

This piece of work was accomplished in so thorough a manner that the scourge of smallpox became practically unknown. Since this brutal despot abdicated and since Mexico entered into the era of true democracy and brotherly love, there has been one scourge after another, which took hold of uninoculated unfortunates who grew up during the latter enlightened years, and which have left entire villages as inanimate as did the Boche some of the French areas.

The general situation in Mexico about the year 1883 when General Diaz was President was chaotic and lawless. But he was astute and strong; he knew men. He knew what the moral and psychological attitude of the *pronunciado* and bandit was. He selected in each district a courageous man, one who was known to be a person of quick action with offensive weapons. This individual was ordered to organize a posse, well-mounted and well-armed.

The leader was called a *Juez de Cordada,* or sheriff, with the power of life or death under certain conditions. The *Juez* was notified that lawlessness was out of fashion; he and his posse were to make such affairs unpopular; this was to be effected as quickly as might be; they were to take care not to double-cross the Diaz Government, because then, in their turn, they would receive immediate liquidation. Within three years Mexico was as safe as a church. These *Jueces de Cordada* had been compelled to execute fewer criminals than would have been expected. They had no chance at all if they were caught red-handed, no trial, nor a bail, nor an appeal, nor any other demoralizing refinement enjoyed in the higher civilization. The culprit was shot down when he was captured or while he was fleeing. Sometimes he was made to flee and then shot down.

Terrible? Well, maybe. But they gave a murderer, a ravisher of women, a thief, exactly what he had been giving to others . . . and those others had been helpless, law-abiding citizens.

The enemies of General Diaz used to claim that many innocent men were killed in this way. Diaz answered that "innocent men had no business in that bad company; that the State is greater than the individual; that no State can prosper unless there is internal peace; that internal peace is not possible where there is no discipline."

President Diaz was a man of enormous courage. His courage carried him through a series of years filled with violence and disorder and enabled him to bring peace and advancement. His courage held back the hand of many assassins and would-be assassins. There was a little malcontent in Mexico who was an anarchist, an anti-anything-decent if he could make a dollar from it. He was a paltry thing. More than once this fellow was warned and his anarchistic sheet was suppressed. But he refused to believe in moderation and kindness shown him; therefore, after a while he was escorted to the Rio Grande where he crossed into the United States and established himself in California. Before long, the authorities there had him incarcerated for his anarchistic activities. He was not really even an anarchist. He was simply a dangerous little beast practically demented, but he was the type who inspired other cowardly morons to assassination. After a while he was freed and joined the campaign against General Diaz, who was the president of a nation friendly to the United States; he was, in truth, a friend to the United States of America. Magazines published grossly erroneous articles about all this and a thousand other matters. These were profusely illustrated, and showed among other pictures the floggings of peons. But these floggings were not photographs, they were pen-and-ink drawings.

Among the prominent men who came to the fore during the days that General Diaz was bringing order from chaos was the Honorable Enrique Creel, who was at one time Ambassador to the United States. Mr. Creel was the man who had organized the Banco Minero in Chihuahua and developed it into a strong and useful institution. He was a close personal friend of our family and he used to tell many interesting events concerning the early days in Mexico, when, as a small boy, he would make trips with his father. Those trips were regular journeys by wagon trains in order to secure goods which were used in trading with Mexicans in the interior of the country, Chihuahua being the distributing point.

Mr. Creel had a favorite story about the first pair of little boots his father bought for him—it probably was in San Antonio, Texas, where they were procured. Those little boots had brass toe caps and a red patent leather ornamental piece let into the front along the top. Mr. Creel said that his joy in the possession of them was so very intense that he suffered agonies with chafed toes and heels for many days rather than not keep them on his poor little feet so he could strut about and show them off.

When I was a little boy myself, this always seemed strange to me . . . a man of wide and varied experience and taste, a man who had seen much, accomplished much, possessed much, extracting infinite pleasure in recounting the story about his first pair of boots.

General Porfirio Diaz

Mr. Creel, God rest his soul; was probably one of the most brilliant and able financial brains of his time. He was one of those patriotic and capable Mexican gentlemen with whom General Diaz surrounded himself. They ought to have been left alone to solve the difficulties of their own nation. . . .

XXVIII

Twenty Thousand Dollars Worth of Dirty Clothes

Sandy was not the only cook at Batopilas; among many others we had a Chinaman who ran the general mess at Hacienda San Miguel, he was very dependable. Charlie Sing was as honest as any man could be. He never held out a penny of the money given him with which to purchase supplies and other foods, nor did he allow his "boys" to do it either. He had a harder time trying to keep his fat Indian wife in bounds than he did his helpers. This wife suffered from a horde of relations—or rather Charlie did the suffering—who were properly lazy; they positively hated to think that with the sister married to a source of supply such as Charlie by all the rules ought to have been, they could not eat as often as they wished without humiliating themselves by manual labor. One of them said to Charlie one day when the Chinaman had relieved him of a hunk of beef:

"You *Chino sin verguenza*, starving your wife's people! What right have you to take this meat from me? It does not belong to you, and it will not cost you a cent."

The fat wife became more than poor patient Charlie could put up with; he came to me in grave distress. I called in the worthless devil who was paying his attentions to her at the time, and told him that if would leave for the Hot Country and present himself before Don Pancho Torres with the letter which I would give him, that Don Pancho would pay him fifty dollars in cash. I told him that if he failed to do this, he would undoubtedly be taken up as a vagrant and be put to work on the streets with the other loafers. Thus Charlie was eased from the burden of the lover of his fat, worthless wife.

In gratitude Charlie did particularly well for a long time and there were very few complaints about the eatables. But one morning I went in to breakfast, and found our old foundry-man with one of Charlie's ears in each mighty hand beating a tattoo with Charlie's head on the stone wall.

"Hey, Kauffman! Lay off Charlie, he is the only cook we've got!"

"Boss," said Kauffman solemnly, "look at this!" He shoved me the saucer on which his coffee cup had been resting, in which two enormous cockroaches were lying. "I told this damned Chinaman," continued Kauffman, "that one cockroach would be all right. But I'm damned if I'll stand for two in one cup of coffee."

We went out to the kitchen and slapped Hell out of the three boys there, Charlie's helpers, and promised them more unless they confessed who had done the cockroach trick. The youngest of the kids was a decent enough little rascal; he blurted out,

"Nepomecino did it. Nepomecino is a cousin of the dear wife's lover now for long a time in la Tierra Caliente. For revenge upon Charlie because of his treachery to this dear cousin, the departed lover, Nepomecino did the trick."

The too devoted relative got a licking and lost his easy job as Charlie's helper in the kitchen. Next time I saw him he was packing a *zurron* at Todos Santos Mine . . . and very good it was for him physically, mentally, and morally. The boss there was very just but he would stand for no foolishness. If a man could be made out of Nepomecino the difficult piece of work would be accomplished there in the shortest length of time. Often these healthy boys like Nepomecino, who had a mean streak, also possessed a brand of courage which when it was brought forward largely submerged the mean streak, and the result would be a good workman, endowed with qualifications which we cultivated in the field in which he appeared most apt.

As an example of the lawlessness of the Mexicans which was talked and written about so very much by untraveled-in-that-country individuals as well as the press during the regime of President Porfirio Diaz, there were three American bandits who held up the Mexican Central train one time between the towns of Torreon and Jimenez. The express messenger on the train attempted resistance and was badly wounded, I think he died. His statement was very clear that three men held up the train and he gave a pretty good description of two of them. I think he shot the third man—I cannot exactly recall that fact . . . I have seen a lot of dead men since then. At sixty-odd one's memory is none too lucid on unimportant past events. I am only interested in this because of the reasons indicated, and because from this incident and the efforts of one of the hold-up men to escape, we had a good joke on old Kauffman.

Kauffman was on the road from San Miguel Mine to the Hacienda San Miguel—he had been examining an old steam hoist at the mine—and overtook a man who had evidently come down the trail from El Parral which came into the cart road between the bridge and San Miguel Mine. This individual was very palpably an American of distressing appearance, dilapidated shoes, sore feet, a hungry look, needing a shave the very worst way, to say nothing of some clean clothes. It was an almost unheard-of thing for an American to walk into our district in that condition, but it had occurred a few times. Kauffman's curiosity was aroused. He deduced, from replies to his questions, that the fellow had left El Parral a few months past with saddle-and-pack animals bent on prospecting the Sierra Madres. He had not found a mine and he had not become suddenly astoundingly rich . . . as happens in the movies and fiction magazines. . . .

He had lost himself. He wandered and scrambled about, finally he managed to get onto the El Parral trail *sans* saddle-mule and pack-mule. He had experienced a very bad time. Even though he did have a little money with him, he could find no one from whom he could buy anything to eat till at last he had come upon a pack-train on the way back to El Parral. From these people he had bought some food, and had learned that there was a mining camp two days farther on. The mining camp was, of course, Batopilas.

"Another damned tenderfoot," thought Kauffman. But since hospitality is customary in that part of the world, Kauffman took him to the hacienda after stopping at the store to get him some clean clothes. He had a bath, a shave, and turned out to be a presentable young man about twenty-five or so. His baggage consisted of a little bundle of dirty clothes that he was anxious not to relinquish.

Everybody had about as much to do as he could attend to comfortably, and no one could waste time on the affairs of other people. Therefore, it never occurred to us that this stranger could be in any way connected with the more or less recent hold-up on the Mexican Central Railway. A job at the mill was found for him; he had been working about a month contentedly enough when on a certain Sunday afternoon old Macario, the Chief of Police, came to call on the general manager, with a description of the man who was wanted in connection with the hold-up.

It had been reported that this man had got as far as El Parral in his efforts to leave Mexico, but from there he was supposed to have gone into the mountains. Since there were only a very few trails he could have followed, and these had been searched without results (no passage or arrival of any strangers whose presence was not explainable) the authorities were checking up, and this fellow was the only person unaccounted for. The general manager acceded the reasonableness of Macario's attitude, consequently the young stranger, West we called him, was sent for. Of course, he repeated the first story he had told us, and to give him credit, he never varied it a hair's breadth. This in itself to the mind of a Sherlock Holmes would have damned him. But it seemed to meet the requirements as far as Macario went, and the description the Chief brought of the possible culprit did not tally with West's appearance.

Following the custom, Alex, who was then general manager, ordered some drinks for refreshment, and he passed around some cigars. At this point West asked to be excused to get his pipe because he did not smoke cigars. Macario gave his consent, the drinks arrived; but after a reasonable length of time, West did not.

Alex suggested to Macario that they should go to see what had happened, it all looked pretty phony to him. West's room was at the very end of the long portal of the inner quadrangle, it had a window opening on an alley or passageway from which, via a cobblestone ramp, the upper level of the hacienda could be reached. The room was empty.

A hurried inquiry and a general search were made with no satisfactory outcome. Macario decided that West had made his get-away, and hastened to town to organize a posse, thoroughly convinced that West and the hold-up man were identical. Darkness had fallen at the time when West had gone to get his pipe, therefore, there was practically no hope of being able to do anything about it that same night. But the trails leading from the town were covered.

Supper time at the mess was lively with discussion about the matter; Kauffman was gibed and questioned about how much his split had been; how in the world it could have happened he had not known that West's bundle of dirty clothes which he could not bear to be parted from was worth twenty thousand dollars; whether he was accessory after the fact, and so on *ad nauseam.*

"That twenty thousand dollars," growled Kauffman, "would have finished paying for my ranch out near Guerrero. Old Kramen and me could have laid off and took it easy. Well, that's what you get for doin' a good act."

West was captured after all. He shot a woodsman, who,

inspired by the reward offered, had attacked the bandit with a knife, the poor fool, and for his pains had stopped a .44, thereby acquiring some experience if nothing more. West was held in jail in the town after being brought there from Sinaloa, and he was still there when the town was captured during the Madero revolution. He was released with all the other prisoners, which was the usual manner of doing such things. I believe that he joined with the revolutionists and became quite a leader in a small way. But I do not think they ever regained the balance of the hold-up money. Sometimes I doubt that he had it in his possession. With that much cash as a stake in those days, a man of any intelligence whatever and with only a bit of experience, having time to study the maps of the section, would have worked out some scheme by which he could have made himself very hard to catch.

Poor old Kauffman! He was able by and by to develop that ranch of theirs to a point where the production of potatoes alone was giving them a fine annual income. They were shipping several carloads yearly and there was a never-failing good market for potatoes. When Pancho Villa got busy, old Kramen was murdered and every single thing they owned that could be carried off was confiscated, and all the buildings were burned. A nice place completely destroyed.

Kauffman was a splendid man, exceedingly accomplished in his profession of foundry-man. When he was in a hurry he would use a three thousand six hundred pound engine bed plate as a model from which to cast a duplicate. Not every foundry-man when he has no time to make a wood model can swing the engine bed plate up on his crane and then lower it into the mold, and sneak it out delicately enough to leave a perfect matrix in which to pour his molten iron.

The mess hall and kitchen for the foreign and native employees was reached by a wide *cantera* stone stairway which spanned the aqueduct from the second level or terrace to the first level below; this was at the up-river side of the Hacienda

San Miguel. There were three dining-rooms, that nearest the stairway was for the foreign employees, next came one for the natives; then the several rooms used by the cook and his waiters and helpers. There were generally from eight to ten foreigners who held different positions of trust—usually one mining engineer whose duty it was to keep the maps up to date and to make all necessary mine surveys; the bookkeepers, a senior and a junior; an assayer (chemist); head machinist; and last but not least old Kauffman.

I will call over some of these able men as best I can with my jaded memory. They were a self-reliant aggregation; they had hard jobs, at time very galling, but said little about it . . . if it were possible to do it they succeeded. Many of those with whom we first worked have passed on. I will wager that in heaven where they are enjoying their just rewards they are frequently called on to help out, for they were grand men in emergencies.

After the Jack Ogden era there was Frank English, a damned fine bookkeeper. Frank was like a bank, there was no use maintaining that he was wrong when you were in the red at the end of the month . . . he always had the irrefutable and damnable evidence of the necessary I.O.U.s to back up his statements.

In the same essential profession in rotation there was Jem Smith, a good fellow, now dead. Then came White, who for many years has held a prominent position with the Government doing things with figures to make the careless watch their steps. Gilbert Geiselhart was the boss machinist who owned the bull terrier, of whom more anon. Old Bill Schwender, who first assisted in making the gasoline locomotive and gasoline hoists at the Weber works in Kansas City, later came down to Batopilas to set them up and to show us how to make them start and go. If Bill played out after a brief spell of fourteen hours at work, all he needed to freshen him up for another go was a couple of quart bottles of beer and a few sardine sandwiches.

Then there was George Bryan who came to Camp to secure lucrative employment the same time Clarence Foreman and Gus Morehouse did, when the Topolobampo colonization scheme died. And Louis Mohun who got fed up on Batopilas and never came back there from a vacation in the States. He is buried somewhere in Las Filipinas. He did a hitch in the United States Infantry, achieved a lot of good and valiant fighting. He was promoted to a Lieutenancy from the ranks as a result of his being a real *hombre* on the field of battle. He was bumped off by a Moro out in the jungles, and his commission waiting for him in Manila! But so men are, and there's no use in regretting anything . . . much. *Le tocó a Luis*, it touched Louis, and his mortal remains are a long way from where he was born.

Ned Moffat, chemist and assayer, whom I first met when I was at Columbia University in 1896, with whose assistance I absorbed large quantities of Wurtzburger on Hammerstein's old roof and other places of joyful memory. Skipper Wells, Coyote Yeandall, but Coyote, the darned rascal, should have been enumerated along with Frank English, for he did a bit of that work himself. He came to Batopilas via Sewanee. Old Bill Kirby-Smith, bless his heart, was there for so many years, had become so much a part of the landscape I nearly forgot him, but he was generally at one of the mines and wound up at San Miguel as superintendent. Later on Bill's brother, Eph, and Louie Loew came along.

More than once in my life have I noted that as one becomes very familiar with a landscape even a prominent feature by its very familiarity may be not seen at times. Among individuals in Batopilas whose life was so identified with the affairs of the company was E. W. A. Jorgensen. He was Assistant Secretary, Secretary, Chief Accountant, and lastly, if

I am not mistaken, Treasurer. He was a great deal more which cannot be set down here. For he is a modest man. He spent by far the greater part of his time in the New York office of the company, but there were many trips that he made to Batopilas. I myself have done many journeys with him across the Sierras. That is a pretty darned good way to size a man up. "Jorgie" was one of the nicest traveling companions that any man ever had.

The ones I have touched upon were the old standbys for years, but there were others who did not last so long. There was Todd who hit camp as engineer, straight from the Scottish Highland via the Royal School of Mines, a true Scot. I will never forget the day that haggis in cans which he had ordered especially from Scotland came into camp. They had spent many months of delay en route and had probably reposed during part of them in a very hot warehouse in the city of Vera Cruz. As a special treat Todd had some of the cans of this delectable haggis opened one night at mess. . . .

Only one other time did the mess hall go through an experience of so foul a stench, this was when Geiselhart's white bull terrier discovered and attacked a skunk which had secreted himself in the fireplace behind some green pine branches on a warm Christmas day.

It was fortunate that Christmas dinner had not yet been served. We bribed a *mozo* to take the dog off and scrub her with hot water and soap using a deodorizer supplied by the doctor. After the room had been whitewashed twice, the brick floor scraped, and every particle of the woodwork washed and repainted, we could return to it as a dining-room.

I have in mind one particular late afternoon when Todd and I were hurrying back to Batopilas from some prospects we had been examining near the San Miguel group. We made our camp for the night in a nice grassy plot on the bank of a small stream after a hot and very annoying day. We had killed some good-size rattlesnakes during the day, and I hate snakes. So did Todd. After supper we spread our blankets and went to sleep. I was dead to the world, when "Bang! Bang!" I came up standing to find Todd doing likewise, but he had his smoking gun in his hand and was blazing away at a spot a few feet from us where my saddle, now plainly visible in the brilliant moonlight, lay with bridle and halter rope over the pommel.

"There it is," cried Todd. "Hells bells, I do believe I missed it!"

"Missed what?"

"The snake, the damned rattler. I see his head sticking out from under the edge of your saddle."

I walked over with quiet tread, and found with burning indignation that what Todd had taken for the head of a rattlesnake was the elaborate, costly and finely made button on the end of my horsehair halter rope. I also discovered that he had placed three .44's in my favorite saddle-tree.

"Just for this, I'm damned if we don't saddle up right now and light out of here to utilize this moon to illumine our way home. *Vamonos, muchachos, aparejan las mulas, vamos de aqui, o no sea este Escocez nos afusila como tigres o elefantes!* Let us go from here, or maybe this Scotchman will shoot us for tigers or elephants."

XXIX

The Kid and Villa

The Kansas City, Mexico and Orient Railroad, building out in a generally southwesterly direction, reached a locality where they established a station which was christened Creel. This made it possible for our company to shorten the journey by pack-train to three days instead of five, and did away with the wagon haul entirely. To accomplish this, and to gain a saving in distance, to avoid very broken country, it was necessary to construct a small suspension bridge across the Eurique River, where this sometimes turbulent stream had been kind enough to cut for itself a narrow gash with nearly perpendicular sides in the rocks of the Sierras. Here by utilizing wire cables (I think it was my brother-in-law, Walter Brodie, who pulled it from his wonderful hat of experience), a very creditable suspension bridge was constructed. As far as I know it is still in use. This must have been somewhere around 1907 or 1908. Walter was more or less of a magician when it came to difficult mining engineering problems; just as Ned Quintard was a wizard at solving problems of how to make rebellious ores yield their values, of telling you just what ore values really were.

The suspension bridge was put together with some of the cable we had taken into camp on the back of the noble and patient mule. There it was given a happier life in the good fresh air and sunshine, instead of being wound and unwound in a dark, deep mine shaft where it was all daubed up at frequent intervals and made sticky by having a combination of tallow and resin smeared on it to keep out the rust and to prevent wear and tear. The wire rope was stretched and securely anchored, a good plank floor was laid, and the sides were boarded up a sufficient height to keep the mules from looking over the edge, realizing where they were, and going over in giddiness and fright.

The Kansas City, Mexico and Orient Railroad was a good idea, but it was never completed for reasons that we will not go into here, because what I would say would undoubtedly be contradicted. I made a trip one time, when that railroad was in process of construction, on a private train filled with prospective investors in the securities of the road. The line had been completed into Texas as far as San Angelo. San Angelo was formerly known as Fort Concho.

In the club car they had an informal reception when we were at San Angelo, at which function were one or two of the oldest inhabitants. One of these, among his many comments, said:

"Yes, gentlemen, I don't think you will find many persons now living who were here before I came to this place and went into the ranching business. I came here with my brother in 1883."

"Well, sir," replied Mr. Stillwell, "there is a man right in front of you who was here three years before you came to old Fort Concho."

"Ha," grunted the oldest inhabitant dubiously, "I sure would like to meet up with him."

"You have," continued Mr. Stillwell, pointing to me. "That is he over there," and he went on to explain.

"Well, and so you are one of that family. Well, well, that is something to talk about after all these years! Why, when we first came here, they was talkin' about those folks from Washington, D. C., who went through here on their way into Mexico. I say, that was some trip for a tenderfoot to make, and I sure always did admire that man. He had plenty of courage to take a family across this country in them days. I am proud to know you, young feller. Say, Mr. Stillwell, why don't you take your crowd down into Mexico that way now? I'll rustle you the teams and damned if I won't go too!"

There were no seconds to the idea.

We drove all over the small settlement out to the old fort buildings, which were still standing, although not in use. In imagination I once more saw the three-year-old Kid with his small stick on his little shoulder trudging toward fancied destruction while the marching troops advanced to pass in review before their commanding officer on old Fort Concho parade ground.

Now the Kid had grown to impressive size, abilities in all his endeavors, and respected manhood. He had a meeting with Pancho Villa at Carretas when this worthless one was living on that section of the country, and developing into the grand patriot that newspapers and others who knew nothing whatsoever about Villa or his misdeeds were building up into a great hero.

Francisco Villa knew quite well who our tribe was, and he knew my young brother. The Kid was a competent horseman, pistol and rifle shot; he stood six feet three in his stockings; weighed one hundred and ninety pounds. There was nothing sticking to his ribs but the very hardest and most efficient and experienced muscle; he possessed a pleasing personality. The Mexicans who knew and worked for him loved him with a devotion not frequently met with. In fact, his position in the estimation of the inhabitants of the sections where his work placed him was one rarely found and much to be desired.

After the Madero revolution started, Conness kept some work going in an effort to develop several small properties out in the vicinity of the old Cusi district. Ojos Azules was in this district. With the advancement of the revolution, the size and importance of the internationally known Pancho Villa swelled with much rapidity. His increase in magnitude and undeniable presence as something actual on the landscape may well be compared to that of a fat mule who has died on the trail, and been subjected to the hot sun of the locality. You soon can see him and smell him from a great distance.

The Kid was making his way into Chihuahua; he rode into Carretas one afternoon accompanied by a couple of his *mozos* and a pack animal or two. He had been there only long enough to wash his face and sit down to supper, when into the town rode Villa with a big bunch of followers. Villa swaggered into the main room of the *meson* where Conness was eating his supper and sat down opposite to him. He had had a few drinks and was bombastic and offensive in elaborating upon the fact that "he was a big mad bull who gored whom he pleased and had to account to no one for his goring." The Kid is a very quiet, a very silent man . . . this kind is the most dangerous.

Villa hated all *gringos*, but he knew exactly who this *gringo* was and he also had no particular amount of guts. When he performed his own shooting and killing, which was seldom enough, he needed all the breaks so that this pastime would be a perfectly safe one to be indulged in by Don Francisco Villa.

The Kid looked him steadily in the eye; that same Kid had a damned mean eye to look into when he felt that way. He then serenely informed Villa that if Pancho started anything it would, of course, result in his, the Kid's, being killed; but he was positive that Villa was aware of the fact that he himself would go first, and that the news of the Kid's death would be transmitted to Villa by some of Villa's many followers in Hell or wherever it was he would be residing when dead. Villa knew that the Kid was right, that when it came to a draw this *gringo* had him beat a mile, so he decided to laugh it off and be a good friendly boy scout. The Kid went on to Chihuahua that night.

The Silver Magnet

I have jumped forward at least thirty-five years or more from the time when you first met the Kid. About three years after you first saw him on the parade ground at old Fort Concho in Texas, one day when my uncle reached Chihuahua with the bullion train from Batopilas, there appeared in that city a tall, nice-mannered, snappy colored man. He approached my uncle and asked for a job at the mines; he knew about mules and horses, spoke some Spanish, was in every way a promising possibility. Out he went to the Sierras. Charlie Robinson (you see I told you there would be more about Charlie later on) proved to be willing, a truly desirable character, and was soon established as night watchman at the main entrance to the Hacienda San Miguel.

Pretty soon Charlie married a seamstress who sewed for our family and in due time two boys and two girls were born. The older boy went to school in the town and became a good mechanic, carpenter, and electrician. The younger boy went to school, too, at times and developed into an errand boy for me at San Miguel Mine. He had his share of paddlings, well-deserved ones they were, by the Yard Boss. In the course of time the soldierly father of these two boys died after a good, faithful, worth-while service; under his dark skin Charlie Robinson was a very fine character.

Came the revolution. It was impossible for the company to continue to operate the mines, because it was impossible for us to get in supplies. Villa attended well to that. This rich property after all those years of very profitable operation shut down and some fifteen hundred men were thrown out of work. That is exactly what Villa wanted. It enabled him to recruit his army to a greater number and that with men from the mining camps . . . a much higher type of man than those he had hitherto been able to obtain.

Among the element thus recruited was this youngest son of Charlie Robinson. We will call him Roberto. By his enterprise, dash and soldierly bearing (inherited from his father;

Charlie had been in the 10th U. S. Cavlary, a non-commissioned officer), Roberto was soon promoted to a Lieutenancy. He was in command of a detachment of men rustling mules and horses for Pancho Villa out in the Zulouaga Ranch section, with headquarters at the splendid old ranch house.

Now, it so happened that my kid brother succeeded in getting his wife out of Mexico and across the border into safety. He decided to return to that section of Mexico to do a little development work in a quiet way on what looked like some good silver-lead stringers. Of course, he was taking a chance, he knew this. He knew that very soon he would have to leave Mexico for good and he thought he would do enough prospecting work on this piece of property so that when he did have to flee across the border to wait for the Mexican situation to clear up, he would know whether these leads were worth keeping in mind for the future or not.

The veins looked good. He had just about made up his mind to stop work, cover under, and light out for the border, some three hundred miles north of him, when his herder came into the camp scared to death. The herder told him that his horses and mules had been grabbed by some of the revolutionists, and herded off to the Hacienda Zulouaga where the bandits had their *remuda* headquarters.

Now, let us jump from Conness' prospecting camp to the *Remuda* Station of Villa's *revolucionarios* at the gate of the Hacienda Zulouaga. There we find Lieutenant Roberto inspecting a group of recently brought-in animals. Among them standing out very conspicously from all the others were a long-maned, jet-black stallion and a couple of bay riding mules.

"Where did you get those?" asked Roberto with an interested expression.

"From a *gringo's* camp about fifteen miles from here."

"Ah, ha," said Lieutenant Roberto to himself. "Those are

the animals belonging to one of my old *Patroncitos* from whom I and my family for many years received nothing but kindness." He turned to the herder: "Put those animals aside by themselves in the small calf-corral. When this has been done, come back here to me."

These instructions were carried out, and when the herder and Lieutenant Roberto were alone, the following orders were given;

"Well after dark, you and one of your men are to take those animals back to the place where you found them. You are to say to my old *Patron* that I return these animals to him so that he can ride north *at once* on the pressing business that he knows of."

Not a bad story when you realize that it was made possible because an excellent colored soldier in self-defense had been forced to kill an undesirable character, a murderer, away back in about 1886 near a United States Army Post close to the Texas border. The sheriff of the place had been glad enough to be relieved of the necessity of shooting, or being shot by, the murderer. After he had a talk with the Post Commander he decided that the culprit, our Charlie Robinson, had gone over the Rio Grande the same night as the killing.

The sheriff anticipated the crossing by only eighteen hours. In no way did he interfere with the crossing, and that is how Pancho Villa happened to have a lieutenant in his army whose father had once worn the chevrons of Uncle Sam's.

Ojos Azules is important to the history of the Pershing expedition into Mexico when our Government sent that able General to chase Pancho Villa. They took good care to limit the chasing to certain areas and they were not kept secret, so that all Villa had to do when too closely pursued was to run over the border line and there to finger his nose at the Pershing expedition. This, of course, made a laughing-stock of our troops and handicapped them so much that any useful result of the affair was patently nullified from the start. At Ojos Azules very early one morning, the Cavalry of General Pershing came upon a camp of the Villa troops, where they made a very satisfactory retaliation for the ambushing and shooting up which these Villa troops had given our cavalry a short time before. This was the high-water mark of the Pershing advance; I have heard that Ojos Azules was out of bounds, but I do not know. However, I do know that the troops of the United States have always given mighty good accounts of themselves wherever they were.

It used to amuse and disgust us to hear and to read what the wise ones of the press said about Francisco Villa. Actually he was a very low type of vagabond bandit. If he ever killed a man personally you can rest assured that the dead man was in a position where he had no show. I feel sure that Villa shot a great many of them in the back. All the romantic stories of his having gone wrong in the beginning because of evil done to his family and of the rape of his sister by an officer of the Rurales Army, are all bunk. He was a rascal before the Madero revolution gave him an opportunity to gather together, as a starter, a group of equally disreputable villains. He was a petty horse-and-mule rustler; this trade was the most dignified one he was ever engaged in, in the way of rascalities. When he had cleaned out the foreign element during his heyday as a military leader he robbed the native shopkeeper, then the native rancher and the native agriculturist, leaving them all utterly destitute. Francisco Villa and a relative of Madero stole and sold into the United States thousands of head of cattle from the Terrazas herds alone, simply leaving no animals at all.

After the second year of his regime, the people at large were so damned sick of the revolution that if they had been given the chance they would have welcomed the interference of anyone who would have relieved them from the depredations of that "patriot." Things got so bad they had to make some sort of deal with him. They gave him a fine ranch near El Parral and a large sum of money to make him be good. He was about through anyway; it was better politics at the time than to bump him off. He had milked that part of his country dry.

Villa graciously accepted the offer; when later he was ambushed and killed as he rode to town in an expensive automobile, he was only paying back for one of the many outrages and killings he had committed or caused to be committed for the sake of personal gain by murder and robbery . . . but always under the guise for military necessity. It was this thief, this murderer, this ravisher of girls, who was honored by the United States Government in detailing a special envoy, a United States Government representative, to accompany him and to see that his murders and depredations were reported in a proper and official light. The moving pictures glorifying Pancho Villa as a persecuted son of the soil did a very grave damage to history for the coming generations.

XXX

Tampico

In 1914 I was offered a position by an oil company in Tampico. This was in the great and beneficent era of freedom from the despotism of General Porfirio Diaz. Our mining operations had perforce ceased.

I needed a job and I needed it badly. The oil company needed men who knew the Spanish language and the Mexican people. Securing oil leases was a complicated racket in the Tampico oil region. The company I was approached by had only one man in the field. I got a wire on Friday in New York where I then was. I accepted the offer, and sailed the next day in a small freighter.

There were precious few sailings to Tampico in those days. Affairs were more than a little unsettled; a part of the confusion could justly be attributed to some of the politically whirring brains in the United States.

The little ship *Camaguey* cast off and went down the river with so great a load of tank iron that the dirty, but fortunately quiet, river was much too high on her sides. There were only six passengers including myself. Being a good sailor I slept well and ate more than my share as I usually do on shipboard. She was pretty steady, being so heavily loaded that she would not or could not rise to the swells. She simply forced her way through them, tunneled them, so to speak. Even though the well-deck was constantly under water and the superstructure continuously soaked by spray, the ship rode smoothly.

The morning we were off Hatteras and blowing hard, the old Scotch engineer, being colossally grouchy, finished off his none too complimentary remarks by the statement that he had been going to sea for thirty years, and this was the first time he had ever sailed in a ——— anvil!

In truth, the *Camaguey* was almost that on the trip to Tampico. There were few ships available and there was a crying need for tank iron. She was loaded full with flat sheets of that metal.

We made a short stop at Vera Cruz, and then on to Tampico. My efforts to find out from conversations in the *cantinas* and other such bureaus of information what was going on were fruitless; the only news was that furnished by the government press, for the town of Vera Cruz was in a state of siege by the *revolucionarios* . . . don't ask me which! You asked questions, you heard answers, you enjoyed, or rather participated in, drinks bought to inspire the emitting of intelligence, and you returned to your ship with a quantity of misinformation. Some of the knowledge obtained did have the merit of vivid imaginary tales and romancings.

One old fellow gave me the startling tidings that the people of the United States had been incensed by the injustices to Mexico perpetrated by the President and Mr. Bryan to such a degree that one had been mobbed and that an attempt had been made on the life of the other! He was much pleased by this exhibition of friendship for Mexico, and we had another drink and cried "*Viva Mexico!*" "*Viva Los Estados Unidos del Norte!*" This was perfectly safe, because we were sure that both of these geographical areas still existed.

I was sorry to take my leave of the gentleman, because I could have afforded to buy a few more *cognacitos;* and it was refreshing to meet a man who had found the philosophical state of mind where, since it was useless to try to talk sense or to act sensibly, the only thing left to do was to drink whatever could be procured, eat when possible, and give an occasional "*Viva Mexico!*" if a Mexican was doing the buying or to substitute the United States if that was what was needed.

Of course, we saw a number of American warships and others lying about; when we came opposite to the mouth of the Panuco and sailed up that stream we also saw two United States gunboats tied to the wharf in Tampico. I asked a gob:

"What the Hell are you doing here?"

"Damned if I know," he replied. "It ain't bad here."

He was conversing in execrable Spanish with a pretty, shapely *mujercita,* she was assisting with yet worse English, and I thought to myself:

"I reckon he's got the right idea. Why worry about fool actions you are not responsible for?"

I reported to the manager of the company in Tampico. He told me my first duty would be to convey more than a hundred thousand dollars in cash for the payroll around to the several camps. This was to be done via the Panuco River and its tributary streams in a fast speed launch. So far he had been unable to obtain passes from the Mexican military commander, who had refused passes to all companies during the past few days.

I was also instructed to go to the lease room or library and to study two or three leases so as to acquaint myself with the forms used. This I did for a few days. I would go at night to the company mess where I slept and ate. This mess consisted of two floors of a big old residence about four blocks from the office of the company. It was about half a block from the corner of the plaza and the soldiers' barracks, El Quartel.

One afternoon as I was walking home, I took a stroll toward the docks and noticed that the American gunboats had cast off and were going down stream. At supper I commented on this fact to the manager, said I thought this disappearance act was a damned funny thing. The feeling against the Americans was growing more torrid daily; if there ever was any need for the presence of the gunboats to protect Americans and American property it would surely be within the next day or two, for things were about ready to break.

"That is exactly what the consul thinks," was his worried answer. "But what makes you think so?"

"Oh, just a hunch, and the information I have acquired in the *cantinas* and cafés where one picks up a bit of stuff if he makes out he is drinking more than he actually is, and if he knows his onions."

"But I feel that you are mistaken," said the company manager, so I let it go at that.

This manager, by the way, did not know a single word of Spanish, he had been at this particular job but a short time past because he had made a good record in the Pennsylvania oil fields. He was a good man, he knew oil. I did not envy him his job with its responsibilities in a situation where no definite stand or policy was taken by his government, nothing but contradictory and vacillating notes being exchanged, and where punishment had been threatened again and yet again with none forthcoming.

It was no wonder that the masses in Tampico were strong in their belief that the "colossus of the north" was yellow and would never take any real action; strong in the belief that they would be absolutely safe in going ahead; confident that if the United States ever did make a military attack Mexico could easily whip the *cobardes*, the cowards.

A Captain of Mexican Infantry told me one night in a *cantina* that they were all set to cross the border at two points, and that it would need no longer than one week to move an army into the City of Washington and take over the United States Government.

He was ignorant and very foolish, of course. He based his calculations on the fact that "Surely you have plainly shown that you are afraid to fight, therefore that the Mexican fighting man could move across your defenseless country at will."

I tried to explain to him that we would undoubtedly fight; that while we might not be able to do much against such ably led troops as those he commanded, not all Mexican troops were so excellent as his. Therefore, there might be considerable trouble in their reaching Washington in so short a time as a week.

"Fight!" he exclaimed. "Fight! *Por Dios*, what makes you think Wilson and Bryan will fight? Have we not insulted you time after time recently? And what do you do? You send ships with orders not to shoot under any provocation. Then you send over another *bombadeo de notas; cartas*, letters! Who was ever hurt by a bit of paper no matter what was written on it?"

The day following the one that saw the gunboats go down the river to join the ships lying off the mouth of the harbor, as I was sitting in the lease room poring over a lease, in came the manager; he informed me hurriedly and a little excitedly:

"They have landed and are fighting in the streets!"

"The Hell they have! This is a small town and I have not heard a single shot."

"Not here," he said, "in Vera Cruz. Here we are at the mercy of these people. The gunboats have gone and the Fleet doubtless had orders to do nothing for our protection. We must get all the office force up to the mess hall and this money too. More than a hundred thousand dollars in cash is in that safe."

By that time, he had the safe open and had extracted two bundles of money done up in sealed packages, unmistakably packages of money. The larger one he handed to me, and the smaller one he gave to one of the other employees. I suggested that a newspaper should be wrapped about these bundles and I did this to mine, trying to make it as careless looking as possible.

"What shall I do with this when I get there?"

"I'll be there before you will. I will put it in the safe we've got there. It is a better one than this one here. I want you to go by the hotel. Go round the block and go into the side entrance. Blank got in last night from up river, tell him to report to the mess hall if trouble comes. We will have a better chance there than we would have at the hotel."

I left. I went round the block as the manager told me and walked back toward the main street and in the side door of the hotel, which was on the corner. The front entrance was on the main street. I don't remember what the name of that main street was now.

I heard shouting, calling:

"*Que mueren los gringos!*" and the like.

I called at the desk and asked for Mr. Blank. He was not in the hotel at the time. I agreed with the manager of the place that it would be better to get his front doors ready to close and to prepare a group of his male guests to be ready for trouble.

In preparation for a possible siege, I went over to the newstand and bought a box of cigars and some cigarettes. I walked to the front door of the hotel carrying my valuable and less valuable packages on my arm. I had my raincoat thrown carelessly over them, and . . . Good Lord! I stepped exactly in the path of a tumultuous mob of angry Mexicans all sweated up and yelling all sorts of death to *gringos* and their ancestors.

There I was, scared as a rabbit. But self-preservation usually crops up strong in the unworthy, and I walked out of that hotel as if nothing in the world were wrong. Halting, I lighted a cigarette and advanced steadily through the very middle of that enraged crowd towards my corner two blocks away.

The mob had not quite yet reached the necessary temperature when any single one of them would have made the first pass at me. Certainly *then* all would have been over in truth. I should have been well messed up at the least, and the company would have been out of a nice piece of money.

But *Dios es grande!* Except for being cursed as a *gringo* and damned for a coward and such abuse, I got through the crowd and up the street to our corner. I am sure I made that half block to our door in one long jump. I don't recall anything about it really. I only remember that I knocked vigorously at the door, made myself known, was admitted, and went up the stairs, feeling the cold sweat running down my back. Tampico was hot as Hell, but I was in a frigid state when I reached our landing.

I went into the main room. I turned over the package of money to the boss and reported that Blank was not at the hotel.

"No, Blank is here," replied the manager of the company. "But where is the fellow with the other bundle of money?"

"How in Hell should I know where he is, you god-damned fool? He left with you, didn't he? Why ask me where he is? To Hell with you and your ——— company, and your lousy money! So long as I did not lose what I was responsible for, I hope you lose a million or two; God knows you can stand it!"

Maybe this was uncalled for. But under certain conditions of heat and pressure I am inclined to pass from the polite to that other state, call it what you may feel inclined. I had never before been subjected to insult where I was in a position of being helpless to retaliate. . . .

Now, I regretted more than ever the confiscation of my two guns when we came into port at Tampico and went through the customs. In all likelihood we were going to be greatly in need of plenty of firearms before morning. Most of the firearms that we had were nearly worthless. You could readily understand our probable need by going to the window and observing the riotous gang, hearing the discharge of firearms and the shouts and screams of "*Que mueren los gringos!*"

That night we kept guard from the windows in the upper

story. I think it was about midnight when it looked very much as if we should have to leave this place, cross the street and get up on the flat roofs of the opposite houses to make a trial at diverting the attention of the mob which was trying to break into the hotel. If there had been no women and children to be endangered in the hotel, we would have welcomed this opportunity to do a little fighting. There was a group of fairly well-armed men inside, they were barricaded efficiently, and there could have been a magnificent clean-up on that rabble, what with the shooting we could have given them from the rear.

The captain of the German ship in the harbor now took a hand and told the commandant of the town that if he did not disperse the mob, he would land his marines and machine guns and clear the streets. This brought immediate action from the Mexican commandant. His cavalry, mounted on their scrawny ponies, dashed out from El Quartel on the corner and into the throng. With the flats of their sabres, and sometimes the edges, the streets were cleared, and the gang scattered to all parts of the city, where they stayed for that night at any rate.

We had the Germans to thank for the prevention of a great loss of life which otherwise would certainly would have been the outcome. Our chagrin and mortification were not lessened by the fact that our own ships with sailors and marines aching to go were at the mouth of the river, but were not permitted to protect their own nationals. The gunboats had been withdrawn lest their presence might incense the Mexicans.

They had already landed that same day and had taken Vera Cruz, but we did not know it then. The troubles in Tampico, where a greater number of Americans were living and working, were directly due to this landing at the other and less important port, indeed I doubt very much if there were any Americans living at Vera Cruz.

All Americans were then ordered to present themselves at

the dock with clean shirts, a change of underwear, toilet articles, nothing more. This order was obeyed. On arrival at the dock we found that a yacht flying the German flag and a tramp steamer flying the British flag or vice versa, were loading the refugees, taking them down the river and out to sea to join the Fleet, where they were being put on the warships.

The unmarried men, particularly the tougher and more experienced ones, were held to the last. They sat on a pile of lumber and other freight on the quay awaiting their turn. We were to act as rear guard if, while the loading was going on, the mob, now collecting across the vacant square opposite to the customhouse, should try to rush the refugees.

Toward the last our Consul arrived at the pier. He had his American flag folded neatly and placed under his arm, was arrayed in immaculate white linen and a Panama hat. He unostentatiously passed through the milling mass of angry Mexicans, behaving in a truly creditable manner, and putting on a very dignified show. Although he was damned and cursed, no one put a hand on him. This gave us a slight compensation for the humiliation we were enduring and had endured. We were glad that under the eyes of foreign sailors he manifested that Americans could carry off any situation as well as anyone, when not under immediate orders which held us back and mortified us.

Our own turn finally came. We were taken down the river, and passed en route very close to a Mexican gunboat. The excited and mostly quite drunken crew had the time of their lives shouting obscenities at us.

Our destination was the old *Cyclops*. We clambered aboard, something over three hundred able-bodied men sore as crabs. We were so very sore that a little while later, when we were boarded by a Sergeant of Marines with a notebook who told us that a ship with the Army Landing Force had been disabled and would not, therefore, be off Tampico in time, and that the Fleet Commander was asking for volunteers for a landing party that very night, he had more than two hundred names just as fast as he could write them.

You can well imagine how thankful we were when we received word that during the afternoon we would be taken off to other ships and be given arms.

We waited; then we waited some more; and then it got dark and we laid down and slept anywhere we could find a space. The old *Cyclops* was a good sturdy collier; but she was a damned poor passenger ship.

Daylight came, and there we were. A swallow of very bad coffee, a spoonful of beans on a slice of bread, was the breakfast. There were no plates and the porridge bowls in which the coffee was served went from hand to hand. This was our ration without change until we landed in Galveston, Texas. It rained every night. There was no shelter for most of us. Were we damp and uncomfortable when we got there!

In Galveston we were denied the privilege of debarking. The United States Government was willing to give us a clean bill of health. They needed the *Cyclops* with her coal to join the fleet at once. But the quarantine officers of "the sovereign State of Texas, By God, sir!" were not willing.

We passed yet another night on that filthy old collier. Came dawn; a couple of cutters pulled over from one of the warships; selected the side away from the shore. Soon a hospital detachment was on board us with a couple of stretchers. In less than thirty minutes they came up from below with these stretchers and something on them covered with sheets.

"What about it?" we asked the medical attendants. "Those men dead?"

"Naw. A couple of 'smokes' with the smallpox."

We gazed at one another in consternation. We decided we were due for quarantine. We hope so certainly; that would beat the old *Cyclops* all to Hell as a place of residence. But no, along about ten o'clock came the Texas authorities and soon we were declared clean. The ship docked; we went ashore.

I had no clothes, little money. But I knew that the manager and those others who had been more fortunate than we had been in the way of ocean transport were in Galveston. I started to make a round of the hotels. By this time, I knew a little about the crowd I had been working for. Therefore, I did not select the best one at which to make my first inquiry. I made a bull's-eye the very first shot.

I went to my room, had my clothes off and was in a tub of hot water in less time than it takes to tell it; sent the bellboy out to get clean underwear and things; while my suit was being scoured, pressed, and promised for the following morning, I ordered up enough supper for two. I put this away and slept for twelve hours—*on a bed.*

At the breakfast table the manager asked me if I would wait over a few days to return to Mexico with him and two other men on the force. The rest of them were going home. Of course, I agreed. A job was a job in those days even at a salary of less than half of what I was accustomed to receiving. Later on that same day, I was advised that we would not be returning to Mexico right away, and that our services were, therefore, no longer required. We would receive a month's pay; our traveling expenses to point of departure would not be paid to us, because we had not completed our two years' contract.

"What about our baggage?"

"That will be returned to you as soon as possible."

But I, for one, was not willing to agree to that. I pointed out to the manager that he would indeed pay my expenses back to point of departure since it was not through my fault that I was back in the States before completion of the two years' contract. That I intended to have the amount of money due me or to demand payment from the President of these United States. Or the Secretary of State. And that I was wiring to Washington about it right away, giving full particulars!!!!

It broke his heart, but he paid our very small and just claim. It was better for him that he did so, because the press was employing large type about this time concerning the whole affair, and reporters dogged one's footsteps.

Months afterwards when at last I received my trunk it contained absolutely nothing at all, and I paid express and other charges to the tune of nineteen dollars. . . .

THE END

Glossary

ABRAZO To embrace a friend upon meeting or departing. A modified embrace, not a lover's death hold

ADOBE Sun-dried mud and straw brick, usually about 15 x 10 x 4 in thickness

ADIOS Good-bye

ALACRAN Scorpion

AMMARRADOR One who ties the knife on a chicken which is to fight

AMIGO Friend

APAREJO Pack-saddle

ARRENDADOR One whose profession is to train a saddle animal to the bit, to make him bridle-wise, to settle him in his gaits, the person who takes the animal over after the buster has done his first part of breaking

ARRIERO Driver of a pack-mule

ARROYO Creek

AZUCAR Sugar

BAILE Ball, dance

BANDERILLEROS Bull-fighters who place the banderillas

BANDERILLA A short rod with small gaff in end. Much adorned by short streamers made of tiger skin or goat skin with hair on, hangs down at back of saddle, on either side. 30 inches long to 36 inches on either side of animal. An adornment, sometimes contains hidden pockets acting as would a saddle-bag

BAQUERILLO Saddle adornment made from tiger skins or long-haired goat skins. Fits over end of saddle-tree behind cantle, like saddle-bags, hangs down on both sides to or a little below line of animal's belly

BARRANCA Gorge

BIEN Well

BIENVENIDA Welcome

BOTANILLA Chamois or buckskin pad that goes around the leg of chicken about to fight and upon which the U of the knife rests

BRIBON Rascal

BURRO Donkey

BUENAS TARDES Good afternoon

BUENOS DIAS Good day

CABALLERO Gentleman
CABRESTO Halter rope, generally horsehair
CAMPO SANTO Cemetery (holy ground)
CANTINA Bar, drinking place
CARABINA Carbine
CARGA Mule load, is used to indicate 300 pounds weight which is the accepted load for slow freight when mules are used. You pay so much freight for a carga of 300 pounds
CARAMBA An exclamation
CERRO Mountain
CHILI Red or green bell pepper. *Chili verde* is green, *chili colorado* is colored (red)
CHIMILCO Small shop or store
CHORRERA A chute in a mine through ores or waste flow from upper to a lower level, also used to indicate a small waterfall
COCHE A carriage or a stagecoach, hack. In some provincial districts a pig is also called a *cochi*
COCHERO A coachman, a stage driver
CON With
CONDENADO One who is condemned, one who is damned. Used as an equivalent of damned, or damn you
CONDUCTOR Boss of a bullion train, mule train, or wagon train
CONDUCTA A train of mules or wagons which transport bullion
CONQUISTADOR Conqueror
CORRAL Fenced-in enclosure in which animals are kept
CORREA Thong of leather
CORRIDA A bull-fight (*corrida de toros*)

DOLOR MUY FUERTE: very strong pain
DON Mister
DOÑA Married lady
DUEÑA Chaperone

EL ESPADA The bull-fighter, he who uses the sword, the head of the group (the quadrilla)

FAJA Sash worn about the waist in place of a belt, it is quite long
FIESTA An occasion of festivity. A day devoted to religious ceremony first and festivities afterwards
FONDA Eating place
FONDERA A woman who runs an eating place
FORMAN A command to form
FORTIN Small fort
FRIJOLE Kidney bean
FUERTE Strong
FUERZA Strength

GATO Cat

GUBERNADOR Governor

GRACIAS Thanks

GRANDE Big, large; principal one

GRINGO Term employed to designate usually a North American or an Englishman, generally the former

GUAYAVA Fruit somewhat resembling a pear in shape but not in taste

HACIENDA The group of buildings at which is established the headquarters of a big ranch. A very large ranch may be made up of several *haciendas*. There are *haciendas de beneficio*, which is used to indicate the place the buildings, the machinery, and so forth where ores are treated and values extracted. You may have a Concession Minera (mining concession) on which you have an *Hacienda de beneficio*

HOJA Leaf, used to designate the roll of corn husks used for cigarettes

HOMBRE Man

HIDALGO A Grandee of old Spain

HUERTAS Gardens, where vegetables are raised, also a small irrigated farm

HUEVOS AL RANCHERO Fried eggs garnished with a sauce made from green peppers, onions, and *chili* powder

INOCENTE An innocent person, also one who is a little lacking in the upper story, quite a young child. A fool

JACAL Small flimsy house, palm roof, sides of interlaced small sticks plastered with mud

JEFE Chief

JEFE POLITICO Political chief, unusually indicates mayor of a town

JEFE POLICIA Chief of Police

JEFATURA Office of the Jefe

MACOUCHIC Very inferior grade of tobacco sold in bulk used in a corn husk cigarette (*Cigarro de hoja*)

MATA Plant or a small tree

MATADOR Bull-fighter who actually kills the bull

MACHETE Heavy broad-bladed cutting knife or sword. Where the country possesses few or no large trees and there is underbrush it takes the place of an axe

MEDIA Half a quart (pint bottle)

MEDIO Half of anything. *Medio real*, half a *real*. A *real* is 12½ cents.

MENUDO Literally small piece; very often used to designate a certain stew made of tripe cut in small pieces with *chili colorado* highly seasoned

MESCAL Distilled colorless liquid made from juice of maguey plant

MESON A hostelry, a place where there is cheer for man or beast.
MESA Literally a table, also used as a tableland
MIL Thousand
MOCHO A person lacking a member, one-armed, one-legged, one-fingered
MONTE Gambling game played with cards. *En El Monte* means out in the bush
MOZO Man-servant
MUCHACHA Girl
MUJER Woman
MUY Very
MUY HOMBRE Very much of a man, a very strong man, a courageous man

NOVIA Woman who is affianced or newly married
NOVIO Man who is affianced or newly married
NAVAJA Small knife, knife used on chickens when they fight

OLLA Earthenware jar, for holding water, etc.

PADRE Father, either of a family or a church
PATRON The big noise in a business, the proprietor
PATRONCITO Small person, the son of the patron
PATIO Central garden in a house that is builded about four sides of same
PECADO Sin
PECADOR Sinner
PELEA DE GALLOS A chicken fight, cocking main
PEQUEÑO Small
PEON Laboring man, not a slave, unskilled laborer
PESO Weight, a dollar
PISO Story of a house, whether first or second floor
PICADOR Bull-fighter who uses lance and rides horse
PISTOLA Pistol
PINTA Term used in indicating certain colors in the ores in the Batopilas district
POBRE Poor
PRONTO Quickly
POR DIOS Y LA SANTA VIRGEN By God and the saintly Virgin
POR MIS GRANDES PECADOS For my great sins
PORTAL A covered porch, a first floor, covered, outside the corridor
POLLO Young chicken
PLAZA Small park or square usually there is a bandstand; also *plaza del mercado,* a market square

QUE What or that
QUE MUEREN LOS GRINGOS Death to the *Gringos*

Glossary

QUADRILLA The group of bull-fighters making up the team for any particular bull-fight

REATA Rope of any material, a rawhide *reata* for roping or the reata for tying on the pack on a mule

REAL Used sometimes to indicate a town. *Vamos al real* Let us go to the town. 12½ cents in money

RIO River

SALA Parlor

SALUD Y PESETAS Health and money

SANTOS Saints

SEÑORITA A young lady

SERAPE Blanket worn in cold weather over the shoulders

SERENATA A serenade; music on the plaza is a *seranata*

SIEMBRA A place where seed has been planted, *Mi siembra,* my garden

SOMBRA Shade

SOMBRERO Hat

SIN VERGUENZA Person without shame, it is a fighting term

TAPADO Covered up, usually indicates a rooster of 6 pounds or more; a heavyweight, free for all. Is not matched for weight

TARANTULA Large black spider, with very long hairy legs, not nice to look at nor to have about the house, poisonous. Plenty of liquor will relieve the bitten person, however, from any permanent disability from its bite

TEQUILLA Liquor distilled, alcoholic drink, plenty of virtue

TOREADOR Bull-fighter

TODOS All

TONTO Stupid one

TORO Bull

TORTILLA Corn cake made from *nestamal,* which is a paste from the grain corn prepared as we prepare the corn for lye-hominy

TUERTO One-eyed

UN One (masculine)

UNA One (feminine)

VAMONOS Let us go, come let us go

VAQUERO Cowboy

VIEJO Old man

VIEJA Old woman

VIVA Live. Exclamation of high approval when it is used followed by the name of the person or thing or place. "Long Life To ——" A female may be *viva,* bright, full of life so may a female mule

VIVO I live. If I am bright while living, I am *vivo.* To tell a person to be *vivo* is to tell him to be very much on the job, also to act promptly or quickly without delay

ZURRON Leather sack carried on the back; capable of containing 150 pounds of ore. Wider at the top than at the bottom, to insure easy emptying of contents. Is used with a necapal, this is a leather strap 4 inches wide which goes around the forehead and extends back on each shoulder and is attached to either side of the zurron about one third distant from the bottom towards the top

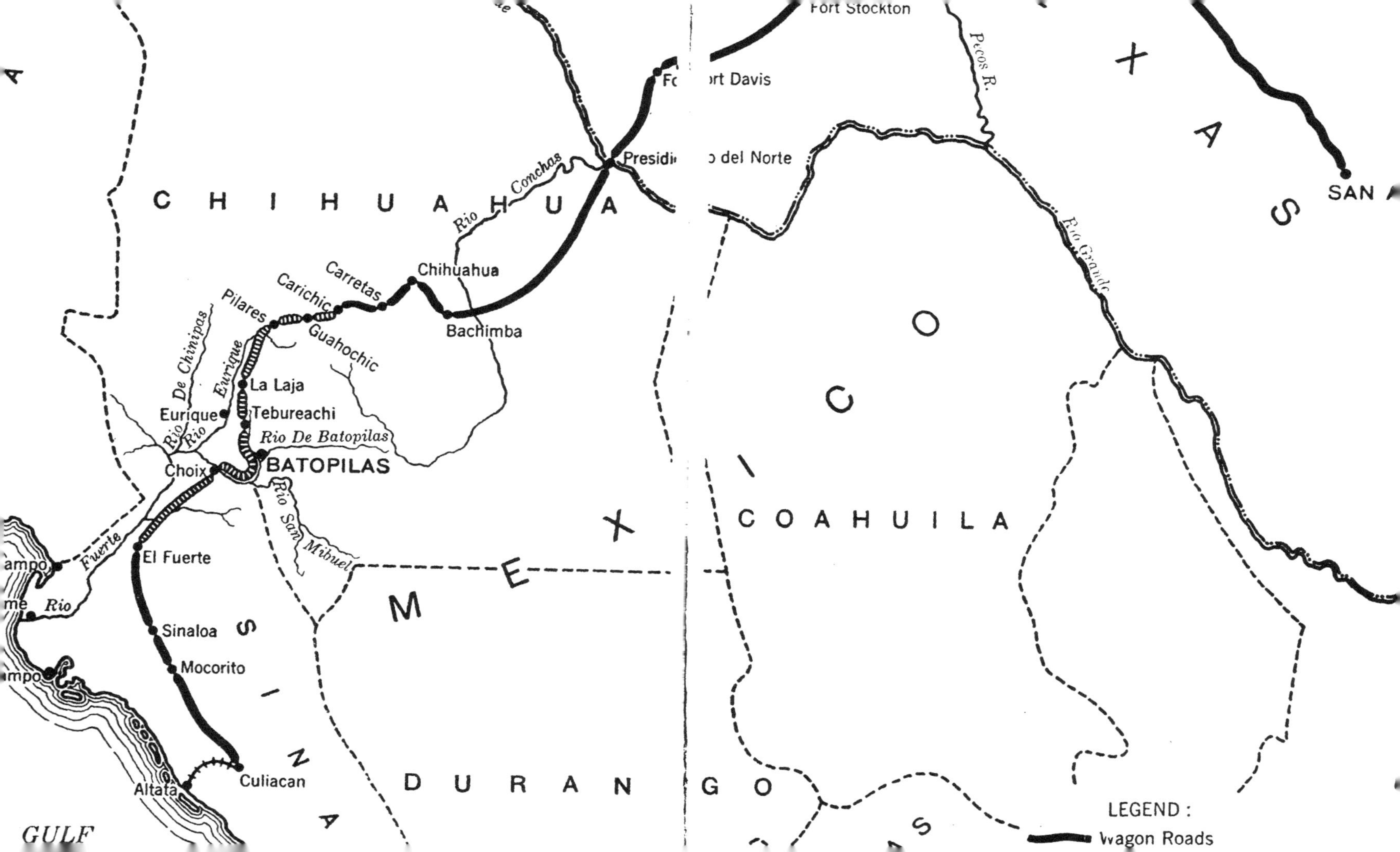
CHIHUAHUA
TEXAS
Fort Stockton
Fort Davis
Pecos R.
Presidio del Norte
Rio Conchas
Rio Grande
SAN
Chihuahua
Bachimba
Carretas
Carichic
Guahochic
Pilares
De Chinipas
Eurique
La Laja
Tebureachi
Rio De Batopilas
BATOPILAS
Rio Rio
Choix
Rio San Mibuel
COAHUILA
MEXICO
El Fuerte
Fuerte
Rio
Sinaloa
Mocorito
Culiacan
Altata
SINALOA
DURANGO
GULF
LEGEND:
Wagon Roads

www.ingramcontent.com/pod-product-compliance
Lightning Source LLC
LaVergne TN
LVHW020535100826
845148LV00010B/1469

9781597405843